# A
# Systems Approach
# to the
# Environmental Analysis
# of
# Pollution Minimization

# A
# Systems Approach
# to the
# Environmental Analysis
# of
# Pollution Minimization

Edited by

# Sven Erik Jørgensen

*Environmental Chemistry*
*Royal Danish School of Pharmacy*
*Copenhagen*

CRC Press
Taylor & Francis Group
Boca Raton London New York

CRC Press is an imprint of the
Taylor & Francis Group, an **informa** business

First published 2002 by Lewis Publishers

Published 2019 by CRC Press
Taylor & Francis Group
6000 Broken Sound Parkway NW, Suite 300
Boca Raton, FL 33487-2742

First issued in paperback 2020

© 2002 by Taylor & Francis Group, LLC
CRC Press is an imprint of Taylor & Francis Group, an Informa business

No claim to original U.S. Government works

ISBN-13: 978-0-367-57913-5 (pbk)
ISBN-13: 978-1-56670-337-6 (hbk)

**Visit the Taylor & Francis Web site at**
**http://www.taylorandfrancis.com**

**and the CRC Press Web site at**
**http://www.crcpress.com**

Library of Congress Card Number 99-30430

---

**Library of Congress Cataloging-in-Publication Data**

A systems approach to the environmental analysis of pollution minimization / edited by Sven Erik Jørgensen.
    p    cm.
Includes bibliographical references and index.
ISBN 1-56670-337-9 (alk. paper)
1. Environmental management. I. Jørgensen, Sven Erik, 1934 –. II. Title: Environmental analysis of pollution minimization.
GE300.S96 1999
658.4'08—dc21

99-30430
CIP

# About the Editor

**Sven Erik Jørgensen** received his masters in chemical engineering at the Technical University of Copenhagen, his Ph.D. at Karlsruhe University, and his Dr. Scient. at the University of Copenhagen. He is now professor in environmental chemistry at the University of Copenhagen, DFH and the Agriculture University of Denmark.

He serves as Editor in Chief of *Ecological Modelling*, chairman of the scientific committee of the International Lake Environment Committee, and has written more than 200 papers in peer-reviewed journals and written or edited more than 39 books in Danish and English on the subjects of environmental management, system ecology, and system approaches to environmental management and ecological modeling.

Dr. Jørgenson has also been awarded visiting professorships at several universities around the world and was named Distinguished Visiting Professor at The Ohio State University in Columbus.

# Contributors

**Rikke Dyndgaard**
Technologisk Institut
International Centre
Taastrup, Denmark

**Michael Hauschild**
Department of Manufacturing Engineering
Technical University of Denmark
Lyngby, Denmark

**Leif Albert Jørgensen**
Danish Statistics
Copenhagen, Denmark

**S. E. Jørgensen**
Environmental Chemistry
Royal Danish School of Pharmacy
Copenhagen, Denmark

**John Kryger**
Technologisk Institut
International Centre
Taastrup, Denmark

**William Mitsch**
School of Natural Resources
The Ohio State University
Columbus, Ohio

**S. N. Nielsen**
Environmental Chemistry
Royal Danish School of Pharmacy
Copenhagen, Denmark

**Hans Schrøder**
Danedi
Helsinge, Denmark

**Bent Halling Sørensen**
Environmental Chemistry
Royal Danish School of Pharmacy
Copenhagen, Denmark

**Henrik Wenzel**
Department of Manufacturing Engineering
Technical University of Denmark
Lyngby, Denmark

# Contents

# 1

## Introduction

S.E. Jørgensen

## CONTENTS

## 1.1   The Basic Concepts of Environmental Management

When the first green wave appeared in the mid and late 1960s, solving the pollution problem was considered feasible. The visible problems were mostly limited to point sources and a comprehensive "end of the pipe technology" was available. It was even seriously discussed in the U.S. to attain what was called "zero discharge" by 1985.

It became clear in the early 1970s that zero discharge would be too expensive, and that we should also rely on the ability of ecosystems to purify themselves. That called for the development of environmental and ecological models to assess the self-purification capacity of ecosystems and to set up emission standards considering the relationship between impacts and effects in the ecosystems.

Meanwhile, it has been disclosed that what we could call the environmental crisis is much more complex than we thought. We could, for instance, remove heavy metals from wastewater, but where could we dispose of the sludge containing the heavy metals? Resource management pointed toward recycling to replace removal. Nonpoint sources of toxic substances and nutrients, chiefly from agriculture, emerged as new threatening environmental problems in the late 1970s. The focus on global problems such as the greenhouse effect and the decomposition of the ozone layer added to the complexity. It was revealed that we use as many as 100,000 chemicals which may threaten the environment due to their more or less toxic effects on plants, animals, humans, and entire ecosystems. In most industrialized countries, comprehensive environmental legislation was introduced to regulate the wide spectrum of pollution sources. Trillions of dollars have been invested in pollution abatement on a global scale, but it seems that at least two or more new problems emerge for each problem that we solve. Our society seems not geared to environmental problems. Or is there, perhaps, another explanation?

Recently, standards for environmental management in industries and green accounting have been introduced. The most widely applied standards for industrial environmental

1-56670-337-9/00/$0.00+$.50
© 2000 by CRC Press LLC

management are the ISO 14000 series. These initiatives attempt to analyze our production systems to find new ways and methods to make production more environmentally friendly. More than 100 countries support the international standards for effective management of environmental impacts.

The goal of ISO 14000 is to evolve a series of generic standards that provide management with a structured mechanism for measuring and managing environmental risks and impacts. Standards have been or are being developed for:

1. Environmental management systems (ISO 14001–14004)
2. Environmental auditing (ISO 14010–14013)
3. Internal reviews (ISO 14014)
4. Environmental site assessments (ISO 14015)
5. Evaluation of environmental + performance (ISO 14031)
6. Product-oriented standards such as environmental labeling, terms, and definitions for self-declaration environmental claims (ISO 14020–14024).
7. Life-cycle assessment, LCA (ISO 14040–14043)

In this volume we will touch on all seven items, but our effort will concentrate on environmental systems. Therefore, we shall go into details about environmental management certification according to ISO 14001. It is based on five steps:

I. Commitment to and creation of an environmental policy
II. Planning, including defining objectives and targets
III. Implementation, including communication and documentation
IV. Monitoring and measuring
V. Reviews and improvement

We shall also consider, from a systems point of view, environmental auditing which is based on a mass and energy balance for an organization or for production. The basic idea is that if we know the flows of matter and energy, we can also identify where we could recycle and where we could reduce loss of matter and energy. Point 7 (above) will also be covered in detail. Life-cycle assessment (LCA) maps the total pollution originating from a product in its lifetime in order to develop products that, to a higher extent, can be recycled, have a longer lifetime, or at least cause less pollution during their lifetime.

In many cases, our production methods have not been charged the actual costs in relation to impact on the environment and the depletion of our resources. A new tax, named the green tax, attempts to charge the real environmental cost on our production. Due to these additional production costs, which are often significant, industries are forced to find new and more environmentally friendly production methods.

Figure 1.1 illustrates how complex environmental management is today. The diagram shows the flow of material (and energy) in the history of a product, from raw materials to final disposal as waste. P indicates emission of point pollutants, and NP covers nonpoint pollution. The number of products in a technological society is not known exactly, but it is probably in the order of $10^7$ to $10^8$. All these products emit pollutants during their production, their transport from producer to user, during their application, and at their final disposal as waste. The core problem in environmental management is: how can we control these pollutants properly? The answer is that we have to use a wide spectrum of methods. Figure 1.1 illustrates where we can use various technologies to control and reduce pollution.

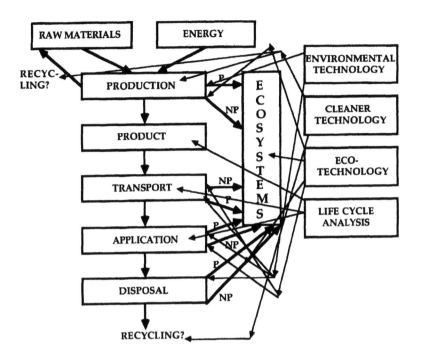

**FIGURE 1.1**

The thick arrows cover mass flows, and the thin arrows symbolize control possibilities. P is point pollution and NP is nonpoint pollution. Production includes both industrial and agricultural production, the latter being mostly responsible for nonpoint pollution. Environmental technology can mainly cope with point pollution, while ecotechnology is needed for the solution of most nonpoint pollution problems, but can also be applied to recover ecosystems. Cleaner technology explores the possibilities of changing present production methods to obtain a reduction in pollution, either by recycling by-products and waste products or by a radical change of production technology. Life cycle analyses are used to find where in the history of a product the pollution actually takes place with the scope to change the product, its transport, or its application.

Four classes of technologies are mentioned: environmental technology, ecotechnology, and cleaner technology including technology resulting from LCA.

Environmental technology offers a wide spectrum of methods that can remove pollutants from water, air, and soil. These methods are particularly applicable to point sources.

Cleaner technology explores the possibilities of recycling by-products or the final waste products or attempting to change the entire production technology to obtain a reduced emission. It attempts to answer the pertinent question: couldn't we produce our product in a more environmentally friendly way? The answer will, to a great extent, be based on environmental risk assessment, LCA, and environmental auditing. The ISO 14000 series and risk reduction techniques are among the most important tools in the application of cleaner technology.

Ecotechnology covers the use of ecosystems to solve pollution problems, including the erection of artificial ecosystems. It also encompasses the technology that is applicable to recover more or less deteriorated ecosystems.

The four classes of technology cover a wide spectrum of methods. We have, for instance, many environmental technological methods for coping with different wastewater problems to select the right method or, most often, a combination of methods. A profound knowledge of the applicability of the methods and of the processes and the characteristics of the ecosystem receiving the emission is compulsory.

Environmental legislation and green taxes may be used in addition to the four classes of technology. They are not included in Figure 1.1, because they may in principle be used as regulating instruments in every step of the flow from raw materials and energy to final waste disposal of the used product.

From this short introduction to the environmental management of today and the wide spectrum of methods that can be implemented to solve the environmental problems, we can conclude that environmental management is a very complex issue. A focal environmental problem may be solved by selection of another raw material or energy source, by changing partially or completely the production method, by increasing use of recycling, by selecting the right combination of technological methods taken from any of the four classes of technologies, by changing the properties of the product slightly, by a combination of environmental technology with recovery of the affected ecosystem and so on. The number of possible solutions is enormous, and yet the environmental management strategy attempts to find the optimum solution from an economic–ecological point of view.

There is, of course, no method available that with high certainty can give the optimum solution to an environmental management problem, but it is clear that the probability of finding a close-to-optimum solution is higher the more knowledge the management team has about the problems and the wide spectrum of solution methods.

This volume attempts to present the characteristics of the various (environmental) systems that should be included in an analysis on which an environmental management strategy should be based. A good knowledge of these systems and a good knowledge of systems ecology, environmental modeling, environmental technology, cleaner technology, life cycle analysis, and ecotechnology are needed. One person cannot cover all these topics properly. We used, therefore, the expression "management team" above. A proper knowledge of the methods available to analyze the systems and processes included directly or indirectly in Figure 1.1 would, however, under all circumstances facilitate the selection of an environmental management strategy. It is the aim of this volume to present this knowledge in a unifying approach, namely by analysis of various environmental systems.

## 1.2 Sustainability

*Sustainability* has become one of the buzz words of our time. It is used again and again in the environmental debate — sometimes in the wrong context. It is, therefore, important to give a clear definition in the introduction to avoid misunderstandings later.

The Brundtland report (World Commission on Environment and Development, 1987) produced the following definition: sustainable development is development that meets the needs of the present without compromising the ability of future generations to meet their own needs. This definition has been widely accepted as authoritative (Willers, 1994). Note, however, that it includes no reference to environmental quality, biological integrity, ecosystem health, or biodiversity. We will touch on all four issues in this volume. An economic treatment of the related concept, *sustainable accounting*, is made by Chichilnisky et al. (1998).

Conservation philosophy has been divided into two schools: resourcism and preservationism. They are understood, respectively, as maximum sustained yield of renewable resources and excluding human habitation and economic exploitation from remaining areas of undeveloped nature. These two philosophies of conservation are mutually incompatible. They are both reductive, ignore nonresources, and seem not to give an answer to the core question: how to achieve a sustainable development, although preservationism has been retooled and adapted to conservation biology.

Lemons et al. (1998) are able to give a more down-to-earth solution by formulating the following rules:

A. Output rule: waste emission from a project should be within the assimilative capacity of the local environment to absorb without unacceptable degradation of its future waste absorptive capacity or other important services.

B. Input rule: harvest rates of renewable resources inputs should be within the regenerative capacity of the natural system that generates them, and depletion rates of nonrenewable resource inputs should be equal to the rate at which renewable substitutes are developed by human invention and investment.

Klostermann and Tukker (1998) discuss sustainability based on product innovation and introduce the concept of eco-efficiency, i.e., the reciprocal of the weighted sum of the environmental claims, including ecological impacts and the draw on renewable and nonrenewable resources.

Klostermann and Tukker (1998) illustrate the total environmental impact, $E$, by use of the equation: $E = P \times W \times I$, where $P$ is the population, $W$ is the consumption per capita, and $I$ is the environmental intervention per consumption unit. The idea is that if $P$ increases as expected to twice the present population by the year 2050, and if we allow $W$ to increase by a factor of, let us say, 5 during the next approximately 50 years, then we have to decrease $I$ ten times to maintain the same impact on the environment, which is a prerequisite for a sustainable development. How can we do that? The answer is by increasing the eco-efficiency ten times. It means to emit 10 times less to the environment and use 10 times less resources for one unit of production. Their book shows that it has been possible to increase the eco-efficiency in a number of industries by a factor of 4. So, why shouldn't we be able to increase the eco-efficiency by a factor of 10 during the next 50 years by using many of the system-oriented methods presented in this volume? This is the message presented by Klostermann and Tukker.

The concept of sustainable development will be applied several times in this volume with reference to the definition given above and with reference to the basic ideas presented in this section.

## 1.3 The Concepts of Systems

A system is an ensemble of components that are coordinated to a certain extent to give the system some characteristic properties. For instance, a production system usually consists of machines, conveyor bands, control equipment, the production staff, and so on. These components are all coordinated to give the production system such properties that allow it to produce a given product.

Chapters 3, 4, 10, and 12 deal with ecosystems; Chapters 5 and 6 with industrial production systems; Chapter 7 with agricultural systems; and Chapter 8 with an extremely complex system, namely an entire region or country. Chapter 9 focuses on the system which you can make around a product from cradle to grave, while Chapter 11 looks into whether the structures of the 100,000 chemicals we are using give the chemicals some characteristic environmental properties — in other words, whether they form a kind of system. The two latter systems may seem different from the others but it should be stressed that the structure of the systems does not need to be entirely physical. It is not presumed in the above definition of a system.

How are systems different from nonsystems? Which characteristic properties should we look for, when we are confronted with systems? Have different systems some characteristics in common? We will try to answer these questions in this section.

All systems, and particularly the systems we will deal with in this volume, are characterized by the following properties:

1.  The organization is hierarchical. If we take as an example a production system, it consists of several classes of components, as described above. Each class consists of a number of components. Each component can be separated into parts, which may be separated into subparts, and the subparts consist of molecules, and the molecules of atoms, and the atoms of neutrons, electrons, and protons, and so on.

2.  The components are connected in one way or another (otherwise they wouldn't be coordinated). These connections can be described, for instance, by use of a *network*. If we know the network in detail, including how it operates, we have a good understanding of how the system works as a system.

3.  The coordinations and connections give the system some characteristic (desirable) properties, often called *emerging properties*. In most cases, the emerging properties are evident. This property is the background for the phrase: a system is more than the sum of its parts.

4.  Information is associated with the use of the system. The information for an ecosystem is embodied in the genes. The production system has an engineering or management plan and so on.

5.  The processes within the system and between the system and its environment can be described as energy-, mass-, and information flows. A production system has a clear mass flow from raw material to final product, unfortunately with some additional flows for waste products. Energy flows are needed mostly in the form of fossil fuel, electricity, and so on to realize the processes; and information flows are the management plans, the engineering schemes, drawings, and so on.

6.  The mass and energy conservation principles can be applied to the system. A system obeys the basic laws of nature, such as the law of gravity and the thermodynamic laws.

    Conservation principles are particularly useful to map the flows and processes in systems and get an overview of what is actually going on in the system and how we, by alterations of the flows, could manage the system better. The description of the mass and energy (and perhaps information) flows in the system, together with the application of the conservation principles, will often be summarized in the form of a model. We shall again and again use the conservation principles as a management tool to optimize the system.

7.  Systems are complex, and a model of the system may be a useful tool to survey the complexity. Ecosystem models are widely used in environmental management (Jørgensen, 1994). We shall, however, not treat the development of models as a surveying tool, although models will be mentioned and applied a few times. Those not familiar with the development of models will, however, have no difficulty following the basic idea behind the use of models in these cases.

8.  Most of the systems we shall deal with in this volume are "medium number" systems. These are understood to be systems with a high number of components (in the order of $10^{10}$ to $10^{20}$), but not such a very high number when compared

to the number of gas molecules in a building or the number of elementary particles in a star. Unlike a system with a very high number of components, the components in a medium number system are all different. The gas molecules in a building can be classified as maybe 12 different types of molecules, and all the molecules of the same class are identical. The number of components in an ecosystem is, let us say $10^{18}$ to maybe 10 orders of magnitude smaller than the number of molecules in a building, but all the components in an ecosystem are different because any two organisms belonging to the same species are different.

## 1.4 Outline of this Volume

This volume will focus on the system properties of ecosystems, industrial and agricultural production systems, a country (or a regional system), the system consisting of the life cycle of a product, and the system consisting of the properties of about 100,000 chemicals we are using in industry, agriculture, and in our everyday life. All six systems have the system properties mentioned above.

Before we apply our system approach to the six systems one by one, we shall present in detail our analytical method for our six systems: the application of the conservation principles for mass and energy. This method is presented in the next chapter, and examples are used to illustrate the method in practice. The analysis leads to a picture of where we have mass and energy (perhaps expressed as concentrations of mass or energy per unit of volume or area). The analysis also encompasses a discussion on what these numbers (amounts or concentrations) mean, or expressed slightly differently, how these numbers can be translated into effects. Such translations are often crucial elements in environmental management.

Chapter 3, *System Properties of Ecosystems*, goes into detail on the complexity of ecosystems, which is necessary to understand the reactions of ecosystems. Ecosystems have, in addition to system properties mentioned in Section 1.2, some additional properties originating from the self-regulation and enormous complexity of these systems.

Chapter 4 presents the use of ecological indicators to take the "temperature" of ecosystems. The idea behind ecosystem health is: just as we can indicate the health of a human being by the use of 10 to 20 different indicators (blood pressure, phosphatase enzymes in urine, hemoglobin content in the blood, and so on), couldn't we in a similar way indicate the health of an ecosystem by use of a similar number of important indicators? And which indicators should we select?

Chapter 5 on environmental management systems and Chapter 6 on cleaner technology present an environmental management approach to industrial production systems. The use of ISO 14001, the standard for the development of environmental management systems in industrial production systems, is compulsory in this context. Its application is presented in details. Cleaner technology is one of the major instruments in using the results of the analysis. Green accounting is an analysis, based on mass and energy conservation principles, which is often used as a tool in the development of cleaner technology.

Chapter 7 presents green accounting with reference to an easy-to-use software. The idea is illustrated by an example from agriculture.

Chapter 8 uses the same principles but for an entire region or country. The chapter discusses how such results can be used to set up environmental strategies for the region or country. It is further discussed how to set up environmental statistics for a country as a basis for development of green accounting for an entire country.

Chapter 9 focuses on life cycle analysis, which is also based on the use of mass conservation principles. The chapter outlines how LCA can be performed and gives illustrative examples.

Chapter 10 covers environmental risk assessment (ERA) for chemicals. Details on how the analysis is carried out are given. Knowledge of the dissemination of toxic chemicals is very important for ERA, but since it is based on many interacting processes which may vary spatially and temporally, ERA will, in most cases, require development of a model.

We don't know all the environmentally relevant properties of most of the 100,000 chemicals we are using in such an amount that they may threaten the environment. Chapter 11 presents how QSAR methods can be applied to estimate these parameters in default of literature values.

Chapter 12 focuses on the application of ecotechnology, which is based largely on the use of technology directly in ecosystems. The chapter attempts to answer the following two questions:

1. How can we consider the characteristic properties of ecosystems when we have decided to use ecotechnology?

2. How can we select the most appropriate type of ecotechnology to solve a specific problem?

The answers are based on a number of focal ecosystem principles which will be mentioned in the chapter with reference to Chapter 3.

Chapter 13, the last chapter, will try to summarize the methods and discuss to what extent they are parallel. The scope of this volume is to show that the methods we are using to approach the management problems of all six environmental systems mentioned above are similar, because they all have some basic (system) properties in common. The last chapter will discuss briefly whether this hypothesis is right or wrong.

# References

Chichilnisky, G., Heal, G., and Vercelli, A., 1998 *Sustainability: Dynamics and Uncertainty*, Kluwer Academic Publishers, Dordrecht.

Jørgensen, S.E., 1994, *Fundamentals of Ecological Modelling*, Elsevier, Amsterdam.

Klostermann, J.E.M. and Tukker, A., 1988, *Product Innovation and Eco-Efficiency*, Kluwer Academic Publishers, Dordrecht.

Lemons, J., Westra, L., and Goodland, R., 1998, *Ecological Sustainability and Integrity: Concepts and Approaches*, Kluwer Academic Publishers, Dordrecht.

Willers, B., 1994, Sustainable development: A new world deception. *Conservation Biology*, 8, 1146.

World Commission on Environment and Development (WCED), 1987. *Our Common Future*, Oxford University Press. Oxford.

# 2

# Conservation Principles and Their Application in Environmental Management

S.E. Jørgensen

## CONTENTS

## 2.1  Conservation Principles and Their Implications

Energy and matter are conserved according to basic physical concepts that are valid for all environmental systems. This means that energy and matter are neither created nor destroyed.

The expression "energy *and* matter" is used, because energy can be transformed into matter and matter into energy. The unification of the two concepts is possible by applying Einstein's law:

$$E = m \ c^2 \quad (ML^2T^{-2}), \qquad (2.1)$$

where E is energy, m is mass, and c is the velocity of electromagnetic radiation in a vacuum (= $3 * 10^8$ m sec$^{-1}$). The transformation from matter into energy and vice versa is only of interest for nuclear processes and does not need to be applied to environmental systems. We might, therefore, break the proposition down to two more useful ones, when applied in environmental management:

1. Environmental systems conserve matter
2. Environmental systems conserve energy

The conservation of matter may be expressed mathematically as follows:

$$dm/dt = input - output \quad (MT^{-1}), \qquad (2.2)$$

where m is the total mass of a given system. The increase in mass is equal to the input minus the output. The practical application of the statement requires that a system be defined, which implies that the boundaries of the system, must be defined.

---

Concentration is used instead of mass in most environmental systems:

$$V \, dc/dt = \text{input} - \text{output} \quad (MT^{-1})$$
(2.3)

where V is the volume of the system under consideration and assumed constant.

If the law of mass conservation is used for chemical compounds that can be transformed to other chemical compounds, Equation 2.3 must be changed to:

$$V * dc_i/dt = \text{input} - \text{output} + \text{formation} - \text{transformation} \quad (MT^{-1})$$
(2.4)

This equation may be applied to the general case of a chemostate, where there is a constant inflow Q (m³ per unit of time) with a constant concentration, $c_{in}$, and an outflow equal to the inflow in volume, but with a concentration of $c_i$ = the concentration in the chemostate. If no formation takes place and a first-order kinetics is presumed for the transformation, we get:

$$V * dc_i/dt = Q \, c_{in} - Q \, c_i - V \, k \, c_i \quad (MT^{-1})$$
(2.5)

where k is the first-order kinetic constant. This equation is often used as an approximation to assess the concentration of a toxic substance in an aquatic ecosystem. It has the following pertinent steady-state solution:

$$c_i = Q \, c_{in}/(Q + V \, k) \quad (ML^3T^{-1})$$
(2.6)

The principle of mass conservation is widely used in the class of ecological models called *biogeochemical models*. The equation is set up for a toxic substance or for some relevant elements, e.g., for eutrophication models for C, P, N, and perhaps Si (see Jørgensen, 1994).

For terrestrial ecosystems, mass per unit of area is often applied in the mass conservation equation:

$$A * dm_a/dt = \text{input} - \text{output} + \text{formation} - \text{transformation} \quad (MT^{-1})$$
(2.7)

where A = area and $m_a$ = mass per unit of area.

The Streeter–Phelps model (Streeter and Phelps, 1925) is a classical model of an aquatic ecosystem which is based upon conservation of matter and first-order kinetics. The model uses the following equation:

$$dD/dt + K_a*D = L_0 * K_1*K_T{}^{(T-20)}*e^{-K*t} \quad (ML^{-3}T^{-1})$$
(2.8)

where

$D$ = $C_s - C(t)$
$C_s$ = concentration of oxygen at saturation
$C(t)$ = actual concentration of oxygen
$t$ = time
$K_a$ = reaeration coefficient (dependent on the temperature)
$L_0$ = $BOD_5$ at time = 0
$K_1$ = rate constant for decomposition of biodegradable matter.
$K_T$ = constant of temperature dependence.

**TABLE 2.1**

Biological Magnification

| Trophic Level | Concentration of DDT (mg/kg dry matter) | Magnification |
|---|---|---|
| Water | 0.000003 | 1 |
| Phytoplankton | 0.0005 | 160 |
| Zooplankton | 0.04 | ~ 13,000 |
| Small fish | 0.5 | ~ 167,000 |
| Large fish | 2 | ~ 667,000 |
| Fish-eating birds | 25 | ~ 8,500,000 |

Source: Data from Woodwell, G.M., et al., 1967, DDT residues in an east coast estuary: a case of biological concentration of a persistent insecticide, *Science*, 156, 821.

The equation states that change (decrease) in oxygen concentration + input by reaeration is equal to the oxygen consumed by decomposition of biodegradable organic matter according to a first-order reaction scheme.

The mass flow through a food chain is mapped by the mass conservation principle. The food taken in by one level in the food chain is used for respiration, as wasted and undigested food, for excretion, growth, and reproduction. If the growth and reproduction are considered as the net production, it can be stated that net production = intake of food − respiration − excretion − (wasted + undigested) food.

The ratio of the net production to the intake of food is called the *net efficiency*. The net efficiency depends on several factors, but is often as low as 10 to 20%. Any toxic matter in the food is unlikely to be lost through respiration, and only minor amounts are lost by excretion, because a toxic substance is much less biodegradable than the normal components in food. This being so, the net efficiency of toxic matter is often higher than that of normal food components, and as a result some chemicals, such as chlorinated hydrocarbons (including DDT and PCB), will be magnified in the food chain.

This phenomenon is called *biological magnification* and is illustrated for DDT in Table 2.1. DDT and other chlorinated hydrocarbons have an especially high biological magnification, because they have a very low biodegradability and are only excreted from the body very slowly, due to dissolution in fatty tissue.

These considerations can also explain why toxic substances and particularly pesticide residues observed in fish increase with the increasing weight (age) of the fish (see Figure 2.1). Generally, the accumulation of a toxic substance with age may be calculated by the following equation:

$$dTox/dt = e_1 \, c_1 \, f + e_2 \, c_2 \, V - k \, Tox \tag{2.9}$$

where Tox is the amount or concentration of the toxic compound, $e_1$ is the efficiency with which the toxic substance is taken up via the food, $c_1$ is the concentration of the toxic substance in the food, f is the amount of food consumed per unit of time, $e_2$ is the uptake efficiency via lungs or gills, $c_2$ is the concentration in air or water, V is the volume of air or water used for respiration per unit of time, and k is the excretion coefficient. It is presumed that excretion follows a first-order reaction.

Because man is the last link in the food chain, relatively high DDT concentrations have been observed in human body fat (see Table 2.2).

The application of the mass conservation principle for production systems may be considered simple bookkeeping of all mass flows. It explains why mass flow mapping is often called an *environmental audit*. The basic idea is to assess the quantity and composition of

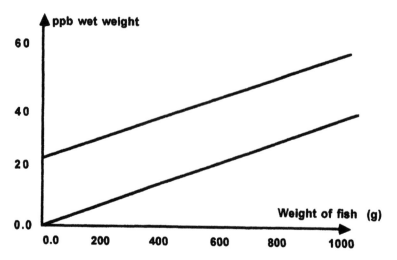

**FIGURE 2.1**

Increase in pesticide residues in fish as the weight of the fish increases. Top line = total residues; bottom line = DDT only. (Adapted from Cox, J.L., 1970, Accumulation of DDT residues in *Trophoturus mexicanus* from the Gulf of California, *Nature*, 227, 192).

**TABLE 2.2**

Concentration of DDT
(mg per kg dry matter)

| | |
|---|---|
| Atmosphere | 0.000004 |
| Rainwater | 0.0002 |
| Atmospheric dust | 0.04 |
| Cultivated soil | 2.0 |
| Freshwater | 0.00001 |
| Seawater | 0.000001 |
| Grass | 0.05 |
| Aquatic macrophytes | 0.01 |
| Phytoplankton | 0.0003 |
| Invertebrates on land | 4.1 |
| Invertebrates in sea | 0.001 |
| Freshwater fish | 2.0 |
| Sea fish | 0.5 |
| Eagles, falcons | 10.0 |
| Swallows | 2.0 |
| Herbivorous mammals | 0.5 |
| Carnivorous mammals | 1.0 |
| Human food, plants | 0.02 |
| Human food, meat | 0.2 |
| Man | 6.0 |

wasted material in order to find alternative solutions to the presently applied production procedure, including, for instance, recycling, which should be considered particularly beneficial from an environmental point of view, because it reduces the utilization of resources and simultaneously reduces emissions to the environment.

The understanding of the principle of conservation of energy, called the first law of thermodynamics, was initiated in 1778 by Rumford. He observed the large quantity of heat that was produced when a hole was bored in metal. Rumford assumed that the mechanical work was converted to heat by friction. He proposed that heat was a type of energy that is produced at the expense of another form of energy, in this case mechanical energy. It was

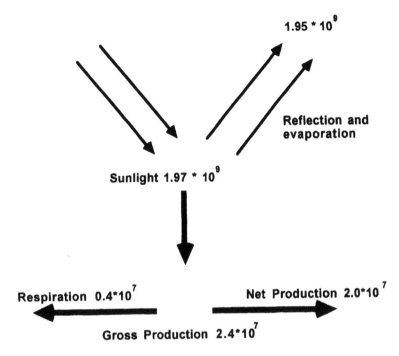

**FIGURE 2.2**
Fate of solar energy incident upon the perennial grass–herb vegetation of an old field community in Michigan. All values in J m$^{-2}$ y$^{-1}$.

left to J.P. Joule in 1843 to develop a mathematical relationship between the quantity of heat produced and the mechanical energy dissipated. Two German physicists, J.R. Mayer and H.L.F. Helmholtz, working separately, showed that when a gas expands the internal energy of the gas decreases in proportion to the amount of work performed. These observations led to the first law of thermodynamics: energy can neither be created nor destroyed.

If the concept of *internal energy*, dU, is introduced:

$$dQ = dU + dW \quad (ML^2T^{-2}) \tag{2.10}$$

where dQ = thermal energy added to the system; dU = increase in internal energy of the system; and dW = mechanical work done by the system on its environment.

The principle of energy conservation can also be expressed in mathematical terms as follows:

U is a state variable, which means that $\int_{1}^{2} dU$ is independent of the pathway $1 \to 2$.

The internal energy, U, includes several forms of energy: mechanical, electrical, chemical, magnetic energy, etc.

The transformation of solar energy to chemical energy by plants conforms to the first law of thermodynamics (see also Figure 2.2): Solar energy assimilated by plants = chemical energy of plant tissue growth + heat energy of respiration.

For the next level in the food chains, the herbivorous animals, the energy balance can also be set up:

$$F = A + UD = G + H + UD, \quad (ML^2T^{-2}) \tag{2.11}$$

**TABLE 2.3**

A. Combustion Heat of Animal Material

| Organism | Species | Heat of Combustion (kcal/ash-free gm) |
|---|---|---|
| Ciliate | *Tetrahymena pyriformis* | −5.938 |
| Hydra | *Hydra littoralis* | −6.034 |
| Green hydra | *Chlorohydra viridissima* | −5.729 |
| Flatworm | *Dugesia tigrina* | −6.286 |
| Terrestrial flatworm | *Bipalium kewense* | −5.684 |
| Aquatic snail | *Succinea ovalis* | −5.415 |
| Brachiopod | *Gottidia pyramidata* | −4.397 |
| Brine shrimp | *Artemia sp. (nauplii)* | −6.737 |
| Cladocera | *Leptodora kindtii* | −5.605 |
| Copepod | *Calanus helgolandicus* | −5.400 |
| Copepod | *Trigriopus californicus* | −5.515 |
| Caddis fly | *Pycnopsyche lepido* | −5.687 |
| Caddis fly | *Pycnopsyche guttifer* | −5.706 |
| Spit bug | *Philenus leucopthalmus* | −6.962 |
| Mite | *Tyroglyphus lintneri* | −5.808 |
| Beetle | *Tenebrio molitor* | −6.314 |
| Guppy | *Lebistes reticulatus* | −5.823 |

B. Energy Values in an *Andropogus virginicus* Old-Field Community in Georgia

| Component | Energy Value (kcal/ash-free gm) |
|---|---|
| Green grass | −4.373 |
| Standing dead vegetation | −4.290 |
| Litter | −4.139 |
| Roots | −4.167 |
| Green herbs | −4.288 |
| Average | −4.251 |

where

F　= food intake converted to energy (Joule)

A　= energy assimilated by the animals

UD = undigested food or the chemical energy of feces

G　= chemical energy of animal growth

H　= the heat energy of respiration.

These considerations pursue the same lines as those mentioned in the context of the application of the mass conservation principle. The conversion of biomass to chemical energy is illustrated in Table 2.3. kJ is obtained by multiplication by the conversion factor 4.18 J/cal. The energy content per g of ash-free organic material is surprisingly uniform, as is illustrated in Table 2.3. It may be compared with the energy content of mineral oil, which is approximately 10 kcal/g or 42 kJ/g. Table 2.3D indicates $\Delta H$, which symbolizes the increase in enthalpy, defined as: $H = U + p*V$.

Biomass can be translated into energy (see Table 2.3), and this is also true of transformations through food chains. This implies that the short food chains of "grain to human" should be preferred to the longer and more wasteful "grain to domestic animal to human."

**TABLE 2.3** (continued)

C. Combustion Heat of Migratory and Nonmigratory Birds

| Sample | Ash-free Material (kcal/gm) | Fat Ratio (% dry weight as fat) |
|---|---|---|
| Fall birds | −8.08 | 71.7 |
| Spring birds | −7.04 | 44.1 |
| Nonmigrants | −6.26 | 21.2 |
| Extracted bird fat | −9.03 | 100.0 |
| Fat extracted: fall birds | −5.47 | 0.0 |
| Fat extracted: spring birds | −5.41 | 0.0 |
| Fat extracted: nonmigrants | −5.44 | 0.0 |

D. Combustion Heat of Components of Biomass

| Material | $\Delta H$ Protein (kcal/gm) | $\Delta H$ Fat (kcal/gm) | $\Delta H$ Carbohydrate (kcal/gm) |
|---|---|---|---|
| Eggs | −5.75 | −9.50 | −3.75 |
| Gelatin | −5.27 | −9.50 | |
| Glycogen | | | −4.19 |
| Meat, fish | −5.65 | −9.50 | |
| Milk | −5.65 | −9.25 | −3.95 |
| Fruits | −5.20 | −9.30 | −4.00 |
| Grain | −5.80 | −9.30 | −4.20 |
| Sucrose | | | −3.95 |
| Glucose | | | −3.75 |
| Mushroom | −5.00 | −9.30 | −4.10 |
| Yeast | −5.00 | −9.30 | −4.20 |

*Source:* Morowitz, H.J., 1968, *Energy Flow in Biology*, Academic Press, New York. With permission.

The problem of food shortage can, however, not be solved so simply, since animals produce proteins with a more favorable amino acid composition for human food (lysine is missing in plant proteins) and eat plants that cannot all be used as human food today. However, food production can, to a certain extent be increased by making the food chains as short as possible.

It is also relevant to set up energy balances for production systems, because they often will disclose an unnecessary waste of energy in the sense that too much heat which cannot be utilized to do work is formed. Energy is of course conserved, which implies that the energy efficiency of any process is 100%. This has provoked the introduction of the thermodynamic concept of *exergy*, which is energy that can do work, to distinguish it from energy that cannot do work, e.g., heat emitted to the environment. So, it seems relevant to set up exergy balances and determine how much exergy is lost by conversion of energy from one form to another. The next section is devoted to the theoretical basis for the introduction of the concept of exergy.

## 2.2 Introduction of the Concept of Exergy

Exergy is defined as the amount of work (= entropy-free energy) a system can perform when it is brought into thermodynamic equilibrium with its environment. Figure 2.3 illustrates the definition. The system is characterized by the extensive state variables S, U, V, N1,

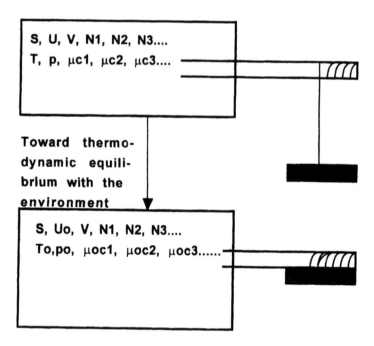

**FIGURE 2.3**
Definition of exergy is shown.

N2, N3…, where S is the entropy, U is the energy, V is the volume, and N1, N2, N3 … are moles of various chemical compounds, and by the intensive state variables, T, p, $\mu c1$, $\mu c2$, $\mu c3$…. The system is coupled to a reservoir, a reference state, by a shaft. The system and the reservoir form a closed system. The reservoir (the environment) is characterized by the intensive state variables $T_o$, $p_o$, $\mu c1o$, $\mu c2o$, $\mu c3o$… and because the system is small compared to the reservoir, the intensive state variables of the reservoir will not be changed by interactions between the system and the reservoir. The system develops toward thermodynamic equilibrium with the reservoir and is simultaneously able to release entropy-free energy to the reservoir. During this process, the volume of the system is constant because the entropy-free energy must be transferred through the shaft only. The entropy is also constant, because the process is an entropy-free energy transfer from the system to the reservoir, but the intensive state variables of the system become equal to the values for the reservoir. The total transfer of entropy-free energy in this case is the exergy of the system. It is seen from this definition that exergy is dependent on the state of the total system (= system + reservoir) and is not entirely dependent on the state of the system. Exergy is therefore not a state variable. In accordance with the first law of thermodynamics, the increase of energy in the reservoir, $\Delta U$, is:

$$\Delta U = U - U_o \qquad (2.12)$$

where $U_o$ is the energy content of the system after the transfer of work to the reservoir has taken place. According to the definition of exergy, Ex, we have:

$$Ex = \Delta U = U - U_o \qquad (2.13)$$

As

$$U = TS - pV + \sum_c \mu_c N_i \qquad (2.14)$$

(see any textbook in thermodynamics), and

$$U_o = T_o S - p_o V + \sum_c \mu_{co} N_i \qquad (2.15)$$

we get the following expression for exergy:

$$Ex = S(T - T_o) - V(p - p_o) + \sum_c (\mu_c - \mu_{co}) N_i \qquad (2.16)$$

As a reservoir, or reference state, we can select the same system but at thermodynamic equilibrium, i.e., all components are inorganic and at the highest oxidation state, if sufficient oxygen is present (nitrogen as nitrate, sulfur as sulfate and so on). The reference state will, in this case, correspond to the ecosystem without life forms and with all chemical energy utilized or as an "inorganic soup." Usually, it implies that we consider $T = T_o$, and $p = p_o$, which means that the exergy becomes equal to the difference between the Gibb's free energy of the system and the reference state, or the chemical energy content + the thermodynamic information (see below) of the system. Notice that the equation above also emphasizes that exergy is dependent on the state of the environment (the reservoir = the reference state), because the exergy of the system is dependent on the intensive state variables of the reservoir.

Notice that exergy is not conserved — only that entropy-free energy is transferred, which implies that the transfer is reversible. In reality, however, all processes are irreversible, which means that exergy is lost (and entropy is produced). Loss of exergy and production of entropy are two different descriptions of the same reality, namely that all processes are irreversible, and we unfortunately always have some loss of energy forms which can do work to energy forms which cannot do work. So, the formulation of the second law of thermodynamics by the use of exergy is: All real processes are irreversible which implies that exergy is lost. Energy is of course conserved by all processes according to the first law of thermodynamics. It is therefore wrong to discuss, as already mentioned briefly, energy efficiency of an energy transfer, because it will always be 100%. The exergy efficiency is of interest, because it will express the ratio of useful energy to total energy which is always less than 100% for real processes. It is interesting to set up an exergy balance for all environmental systems, in addition to an energy balance. Our concern is loss of exergy, because that means that "first-class energy," which can do work, is lost as "second-class energy," which cannot do work.

Exergy seems more useful to apply than entropy to describe the irreversibility of real processes, because it has the same unit as energy and is an energy form, while the definition of entropy is more difficult to relate to concepts associated to our usual description of the reality. In addition, entropy is not clearly defined for far-from-equilibrium systems, particularly for living systems, according to Kay (1984). Finally, it should be mentioned that the self-organizing abilities of systems are strongly dependent on the temperature, as discussed in Jørgensen, 1997. Exergy takes the temperature into consideration, as the definition shows, while entropy doesn't.

Notice also that information contains exergy. Boltzmann (1905) showed that the free energy of the information we actually possess (in contrast to the information we need to describe the system) is $k*I* \ln I$, where I is the information we have about the state of the system (for instance, that the configuration is 1 out of W possible, i.e., W = I), and k is Boltzmann's constant = $1.3803* 10^{-23}$ (J/molecules*deg). It implies that one bit of information has an exergy equal to $k T \ln 2$. Transformation of information from one system to another is often almost an entropy-free energy transfer. If the two systems have different temperatures, the entropy lost by one system is not equal to the entropy gained by the other system, while the exergy lost by the first system is equal to the exergy transferred and equal to the exergy gained by the other system, provided that the transformation is not accompanied by any loss of exergy. In this case, it is obviously more convenient to apply exergy than entropy.

Exergy is closely related to information theory. A high local concentration of a chemical compound, for instance, with a biochemical function that is rare elsewhere, carries exergy *and* information. On the more complex levels, information may still be strongly related to exergy but in more indirect ways. Information is also a convenient measure of physical structure. A certain structure is chosen out of all possible structures and defined within certain tolerance margins (Berry, 1972; Thoma, 1977).

Biological structures maintain and reproduce themselves by transforming energy and information from one form to another. Thus, the exergy of the radiation from the sun is used to build the highly ordered organic compounds. The information laid down in the genetic material is developed and transferred from one generation to the next.

The chromosomes of one human cell have an information storage capacity corresponding to 2 billion K-bytes! When biological materials are used to the benefit of mankind, it is the organic structures and the information contained therein that are advantageous, for instance, when using wood.

Information is of utmost importance for production systems, where the right management can be crucial for the efficiencies of the processes utilized in the system. Green accounting, or environmental audit, is a new method of providing more information about the system.

It can be shown (Evans, 1969) that exergy differences can be reduced to differences of other, better known, thermodynamic potentials (see Jørgensen, 1997), which may facilitate the computations of exergy in some relevant cases.

The exergy of the system measures the contrast — the difference in free energy if there is no difference in pressure, as may be assumed for an ecosystem or an environmental system and its environment — against the surrounding environment. If the system is in equilibrium with the surrounding environment, the exergy is zero.

Since the only way to move systems away from equilibrium is to perform work on them, and since the available work in a system is a measure of the ability, we have to distinguish between the system and its environment or thermodynamic equilibrium. Therefore, it is reasonable to use the available work, i.e., the exergy, as a measure of the distance from thermodynamic equilibrium.

As we know that ecosystems, because of the through-flow of energy, have the tendency to develop away from thermodynamic equilibrium, losing entropy or gaining exergy and information, we can put forward the following proposition of relevance for ecosystems: **Ecosystems attempt to develop toward a higher level of exergy.**

This description makes it pertinent to assess the exergy of ecosystems. It is not possible to measure exergy directly, but it is possible to compute it according to Equation 2.16, if the composition of the ecosystem is known.

If we presume a reference environment that represents the system (ecosystem) at thermodynamic equilibrium, which means that all the components are inorganic at the highest possible oxidation state if sufficient oxygen is present (as much free energy as possible is utilized to do work) and homogeneously distributed in the system (no gradients), the situation illustrated in Figure 2.4 is valid. Because the chemical energy embodied in the organic components and the biological structure contributes far more to the exergy content of the system, there seems to be no reason to assume a (minor) temperature and pressure difference between the system and the reference environment. Under these circumstances, we can calculate the exergy content of the system as coming entirely from the chemical energy:

$$\sum_c (\mu_c - \mu_{co}) N_i$$

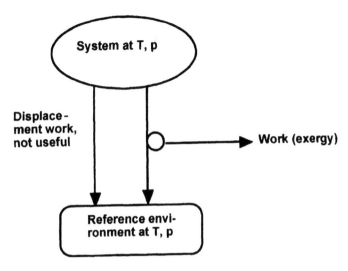

**FIGURE 2.4**
The exergy content of the system is calculated in the text for the system relative to a reference environment of the same system at the same temperature and pressure, but as an inorganic soup with no life, biological structure, information, or organic molecules.

By these calculations, we find the exergy of the system compared with the same system at the same temperature and pressure, but in the form of an inorganic soup without any life, biological structure, information, or organic molecules. As $(\mu_c - \mu_{co})$ can be found from the definition of the chemical potential replacing activities by concentrations, we get the following expressions for the exergy:

$$Ex = RT \sum_{i=0}^{i=n} C_i \ln C_i / C_{i,o} \qquad (2.17)$$

where R is the gas constant, T is the temperature of the environment (and the system; see Figure 2.4), while $C_i$ is the concentration of the *ith* component expressed in a suitable unit, e.g., for phytoplankton in a lake $C_i$ could be expressed as mg/l or as mg/l of a focal nutrient. $C_{i,o}$ is the concentration of the *ith* component at thermodynamic equilibrium, and n is the number of components. $C_{i,o}$ is, of course, a very small concentration (except for i = 0, which is considered to cover the inorganic compounds), corresponding to a very low probability of forming complex organic compounds spontaneously in an inorganic soup at thermodynamic equilibrium. $C_{i,o}$ is even lower for the various organisms, because the probability of forming the organisms is very low with their embodied information, which implies that the genetic code should be correct. It can be shown (Jørgensen et al., 1995; Jørgensen, 1997) that the sum of the exergy of the components in an ecosystem, Ex, with a reference state as shown in Figure 2.4, with approximation, can be calculated as:

$$Ex = \sum_{i=0}^{i=n} \beta_i c_i \qquad (2.18)$$

$\beta$-values for various organisms reflecting their differences in embodied information in the genes can be found from Table 2.4. Here $\beta = 1$ is used for detritus (dead organic matter) without information but with chemical energy (see the chemical energy content of organic

**TABLE 2.4**

Approximate Number of Nonrepetitive Genes

| Organisms | Number of Information Genes | Conversion Factor * |
|---|---|---|
| Detritus | 0 | 1 |
| Minimal cell (Morowitz, 1992) | 470 | 2.6 |
| Bacteria | 600 | 3.0 |
| Algae | 850 | 3.9 |
| Yeast | 2,000 | 6.4 |
| Fungus | 3,000 | 10.2 |
| Sponges | 9,000 | 30 |
| Molds | 9,500 | 32 |
| Plants, trees | 10,000–30,000 | 30–87 |
| Worms | 10,500 | 35 |
| Insects | 10,000–15,000 | 30–46 |
| Jellyfish | 10,000 | 30 |
| Zooplankton | 10,000–15,000 | 30–46 |
| Fish | 100,000–120,000 | 300–370 |
| Birds | 120,000 | 390 |
| Amphibians | 120,000 | 370 |
| Reptiles | 130,000 | 400 |
| Mammals | 140,000 | 430 |
| Human | 250,000 | 740 |

* Based on number of information genes and the exergy content of the organic matter in the various organisms, compared with the exergy contained in detritus. 1 g detritus has 18.7 kJ exergy (energy which can do work).

Sources: Cavalier-Smith, T., 1985, *The Evolution of Genome Size*, Wiley, Chichester; Li, W.H. and Grauer, D., 1991, *Fundamentals of Molecular Evolution*, Sinauer, Sunderland, Massachusetts; Lewin, B., 1994, *Genes V.*, Oxford University Press, Oxford. With permission.

matter in Table 2.3). From these β-values, we get the exergy per unit of volume (or area) in detritus exergy equivalent. If we prefer the unit kJ/unit of volume or area, the exergy found by use of Equation 2.18 should be multiplied by 18.7, which is the average chemical energy content of detritus (compare with the values in Table 4.3).

The total exergy of an ecosystem *cannot* be calculated exactly, because we cannot measure the concentrations of all the components or determine all possible contributions to exergy in an ecosystem. If we calculate the exergy of a fox, for instance, the calculations shown above will only give the contributions coming from the biomass and the information embodied in the genes. What is the contribution from the blood pressure, the sexual hormones, and so on? These properties are at least partially covered by the genes, but is that the entire story? We can calculate the contributions from the dominant components, for instance by the use of a model or measurements, that cover the most essential components for a focal problem.

Exergy calculated by Equation 2.18 has some shortcomings. It is proposed that we consider the exergy found by these calculations as a *relative exergy index*.

1. We account only for the contributions from the organism's biomass and information in the genes. Although these contributions probably are the most important ones, it cannot be completely excluded that other important contributions are omitted.

2. We don't account for the information embodied in the network — the relationships between organisms. The information in the model network that we use to

describe ecosystems is negligible compared with the information in the genes, but we cannot exclude that the real, much more complex network may contribute considerably to the total exergy of a natural ecosystem.

3. We have made approximations in our thermodynamic calculations. They are all indicated in the calculations and are in most cases negligible.

4. We can never know all the components in a natural (complex) ecosystem. Therefore, we will only be able to utilize these calculations to determine exergy indices of our simplified images of ecosystems (models).

Exergy indices are useful, because they have been used successfully as goal functions (orientors) to develop structural dynamic models. The *difference* in exergy by *comparison* of two different structures (species composition) is decisive here. Moreover, exergy computations always give only relative values, because the exergy is calculated relative to the reference system.

As already stressed, the presented calculations do not include the information embodied in the structure of the ecosystem, i.e., the relationship between the various components, which is represented by the network. The information of the network encompasses the information of the components and the relationships of the components. The latter contribution is calculated by Ulanowicz (1991) as a part of the concept of *ascendancy*. In principle, the information embodied in the network should be included in the calculation of the exergy of structural dynamic models, because the network is also dynamically changed (Pahl-Wostl, 1995). It may, however, often be omitted in most dynamic model calculations, because:

1. The contributions from the network relationships of models (not from the components of the network, of course) are minor, compared with the contributions from the components. This is due to the extreme simplifications made in the model compared with the real network, although they attempt to account for the major flows of energy or mass.

2. In most cases, a relative value of the exergy in the form of an exergy index is sufficient to describe the direction of ecosystem development/growth.

3. If the network is changing in addition to the components (the nodes) of the network, it should/could be considered what this change would contribute to the exergy index of the system.

4. The calculations of exergy indices will always be approximations focusing on the most important components with respect to the changes taking place. A model is often used as the basis for these calculations, and a model is always a simplification of the real system. The ecosystem is so complex it is impossible to know all the components.

As previously mentioned, it would be very useful to set up an exergy balance for a production system. It is hardly possible to give general equations that can be applied to all production systems, similar to the set of equations given above for ecosystems. However, some general considerations on the development of exergy balances for production systems could be made. Usually the exergy of information need not be included in the analysis, but the input of first-class energy in the form of fossil fuel, electricity, and other energy forms must be computed and compared with the exergy of the outputs and/or with the loss of exergy in the form of useless heat from the entire production system. Useful heat is heat associated with a temperature higher than that of the environment. It has work capacity, just as heat input for maintenance of a suitable room temperature also contains exergy. The waste heat transferred to the temperature of the environment has no exergy content.

It could be considered whether non-useful heat could be made useful, but it would require that the heat still have a temperature higher than that of the environment.

In principle, can (and should) the exergy balance of each process in a production system be analyzed to find if the exergy input and/or the generation of non-useful heat could be reduced? The exergy efficiencies of many of our production processes are surprisingly low. There has been an increasing interest in reducing the exergy consumption with the increasing exergy (energy) cost and decreasing access to low-cost energy. The exergy consumption of a modern refrigerator is, for instance, only 30 to 40% of that of a 20-year-old refrigerator with the same capacity. By more comprehensive use of exergy balances, it will be possible to indicate how to reduce the overall exergy consumption in the developed countries. If we, on average, could increase the exergy efficiencies from 40% to 60%, it would imply a reduction of our energy consumption and the associated emissions of pollutants by 33.3%.

## 2.3   From Emissions to Effects

A knowledge of emissions is needed to set up mass balances for a system — emissions to the internal or external environment. Equations 2.4 through 2.6 may often be used in the most simple cases, but when the problem gets a little more complex, use of models is almost compulsory, because a good overview and computations of several interacting processes simultaneously is required. It is not the intention in this volume to present modeling as a useful tool in environmental management, as it would be too comprehensive.

A classification of ecotoxicological models should, however, be mentioned, because it illustrates very clearly the thoughts behind modeling in environmental chemistry. Three classes of models may be considered (see also Section 10.3):

A. Fate models, including fugacity models.
B. Ecosystem-specific models of toxic substances.
C. Models of toxic substances in a typical or average ecosystem, which is used as a general representation of all ecosystems of a given type.

Class A models are used to get a rough estimate of where a toxic substance will be found in the environment and at approximately which concentration. This class of ecotoxicological models is useful for comparison of various alternative chemicals. It may be used to answer the following pertinent question: should we prefer chemical X or Y from an environmental point of view? This type of model has been treated comprehensively in D. Mackay's book, *Multimedia Environmental Models.*

Class B models are used when we have to decide on abatement of a specific pollution of a toxic substance, for example when an organic substance is transported from a chemical plant by wastewater to the environment, perhaps after the wastewater has passed a treatment plant. In this case, it is necessary to include the characteristic features of the ecosystems receiving the toxic substance in the model. This type of ecotoxicological model is obviously similar to ecological models, because it must include the same characteristic ecological features as ecological models in general. This class of model is generally more accurate than class A models, but it is also more specific and can rarely be applied for ecosystems other than the one the model was developed for. This class of models is presented in detail in Jørgensen, 1994.

Class C models have recently found an increasing application because many uses of chemicals affect specific types of ecosystems. If we want, for instance to select a pesticide which is most harmless to earthworms, we have to "test" our spectrum of pesticides on a general agricultural ecosystem model with a food chain that includes earthworms. We don't have a specific agricultural system in mind, but rather a very general agricultural ecosystem with average features characteristic of such systems. In this case, we will develop a model with the characteristic features of an average agricultural system. This class of models will be less accurate than class B models but will generally give more accurate and more specifically applicable results than class A models. On the other hand, the results cannot be applied with the same generality as for class A models.

The results of computations or a model are concentrations in different parts of the environment. They have to be evaluated, which implies a "translation" into an effects. This is a very difficult task because of our limited knowledge about the environmental effects of the enormous amounts of chemicals we use in everyday life or in our production systems. In addition, there is a wide spectrum of possible effects and, therefore, many toxicity expressions and indices to cover these effects. The effects include lethality, growth inhibition, carcinogenicity, teratogenicity, mutagenicity, reduction in enzyme activity, narcotic effect, and a number of sublethal effects.

The most important indices to indicate these effects are:

$LC_n$: a calculated concentration which, when administered by the respiratory route, is expected to kill n% of the population of experimental animals during an exposure of a specified duration. Ambient concentration is expressed in milligrams per liter.

$LD_n$: a calculated dose of a chemical substance which is expected to kill n% of a population of experimental animals exposed through a route (often through food uptake) other than respiration. Dose concentration is expressed in milligrams per kilogram of body weight.

$IC_p$: the inhibiting concentration to produce an inhibiting effect of p%.

$EC_q$: the concentration to produce a specified effect on q% of the experimental organisms.

(The Glossary gives a more comprehensive list of toxicity expressions and indices.)

Environmental management is particularly interested in the assessment of safe concentrations, i.e., concentrations where no effects are observed. Terms widely used in environmental management include:

No observed effect concentration (level) — **NOEC (NOEL)**: defined as the highest concentration (level) of a test chemical substance to which the organisms are exposed that does not cause any observed and statistically significant adverse effects on the organisms compared with the controls.

Predicted no effect concentration — **PNEC**: which is the environmental concentration below which it is probable that an unacceptable effect will not occur according to predictions. PNEC is found from NOEC by use of a suitable safety factor (see Chapter 10).

One of the core elements in environmental risk assessment (for further details, see Chapter 10) is a comparison between PNEC and predicted environmental concentrations (**PEC**), which is the concentration of a chemical in the environment, calculated primary by use of models on the basis of available information about its properties and application pattern.

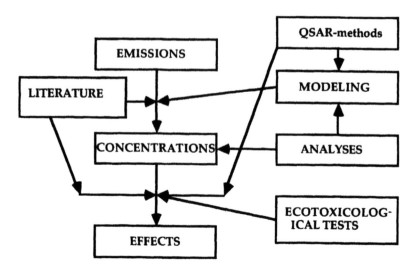

**FIGURE 2.5**

Environmental management of emissions is concerned with a "translation" of the emissions to concentrations in the environment and a "translation" of these concentrations to effects. This requires the use of models, analyses, and ecotoxicological tests. The "translations" require knowledge of parameter values, which may be found in the literature or may have to be estimated by use of QSAR methods.

For a working environment, it is usual to apply a maximum allowable concentration, **MAC**: a value in accordance with environmental legislation. It is often dependent on time. The applied unit is usually mg/m³. It is assessed on the basis of the best possible knowledge about the effect of chemicals. It is considered a safe concentration by a daily exposure of eight hours.

Since our food and drinking water may be contaminated, it is necessary to assess an "acceptable daily intake" (**ADI**), which is the estimate of the amount of a considered substance in food or drinking water that can be ingested over a lifetime by humans without appreciable health risk. ADI is also used for food additives. The applied unit is usually mg/kg body weight.

Our knowledge of the physical–chemical, biological (e.g., biological concentration factors, biodegradation and biological magnification factors), toxicological, and ecotoxicological properties of the applied chemicals is unfortunately very limited. We are using about 100,000 chemicals in such an amount that they may threaten the environment, but we have perhaps on the order of 1% or less of the knowledge needed to make a complete evaluation of the risk of these chemicals. These chemicals comprise everyday products including detergents and cosmetics, industrial chemicals, fertilizers and pesticides used in agriculture, a number of intermediate chemicals which are used as raw materials in other industries, drugs, and additives in food. We have either to rely on estimates of these properties on the basis of the structure of the chemicals (see Chapter 11), or to apply a suitable and feasible selection of the wide range of available toxicity and ecotoxicity tests.

Figure 2.5 summarizes the considerations of how to combine knowledge about emissions and effects, computations, analyses, and effects tests to evaluate the effects of emissions. The objective is to assess the acceptable level of the emission from an environmental point of view. This is particularly relevant for emissions of toxic substances to ecosystems, but the same procedure is in principle applicable for production systems where the concentrations in the working environment are pertinent.

As seen in the figure the emission is "translated" into a concentration by use of computations and models. The concentrations are found by application of analytical chemistry to calibrate or validate the model in some cases and to increase the certainty of an application of the model in other cases. It is not feasible to analyze all possible parts of the environment. It would be an impossible task. The required number of analyses would be much too high. By a suitable combination of analyses and modeling, a high certainty can be achieved on the basis of a feasible number of analyses. The development of a model requires knowledge of a number of parameters for the considered chemicals. A handbook (see Jørgensen et al., 1991) or other relevant literature is needed to obtain these parameters, or it is necessary, as shown in the figure, to apply QSAR methods; see also Chapter 11. The "translation" of the concentration to effects requires knowledge of biological, toxicological, and ecotoxicological parameters, which may be found in handbooks or other relevant literature, by use of QSAR methods, or on the basis of laboratory tests, as described above.

# References

Berry, S., 1972, *Bull. Atomic. Sci.*, 9, 8.

Boltzmann, L., 1905, The second law of thermodynamics. *Populare Schriften. Essay No. 3.* (address to Imperial Academy of Science in 1886). Reprinted in English in *Theoretical Physics and Philosophical Problems, Selected Writings of L. Boltzmann.* D. Reidel, Dordrecht.

Cavalier-Smith, T., 1985, *The Evolution of Genome Size*, Wiley, Chichester.

Cox, J.L., 1970, Accumulation of DDT residues in *Trophoturus mexicanus* from the Gulf of California, *Nature*, 227, 192.

Evans, R.B., 1969, *A Proof that Essergy Is the Only Consistent Measure of Work*, thesis, Dartmouth College, Hanover, NH.

Jørgensen, S.E., 1994, *Fundamentals of Ecological Modelling*, Elsevier, Amsterdam.

Jørgensen, S.E., 1997, *Integration of Ecosystem Theories: A Pattern*, Second edition. Kluwer, Dordrecht.

Jørgensen, S.E., Nielsen, S.N., and Jørgensen, L.A., 1991, *Handbook of Ecological Parameters and Ecotoxicology*, Elsevier, Amsterdam.

Jørgensen, S.E., Nielsen, S.N., and Mejer H.F., 1995, Emergy, environ, exergy and ecological modelling, *Ecol. Modelling*, 77, 99.

Kay, J., 1984, *Self Organization in Living Systems*, thesis, System Design Engineering, University of Waterloo, Ontario, Canada.

Lewin, B., 1994, *Genes V.*, Oxford University Press, Oxford.

Li, W.H. and Grauer, D., 1991, *Fundamentals of Molecular Evolution*, Sinauer, Sunderland, Massachusetts.

MacKay, D., 1991, *Multimedia Environment Models*, Lewis Publishers, Boca Raton, FL.

Morowitz, H.J., 1968, *Energy Flow in Biology*, Academic Press, New York.

Nielsen, S.N., 1992, *Application of Maximum Exergy in Structural Dynamic Models*, Ph.D. Thesis, DFH, Institute A, Section of Environmental Chemistry, Copenhagen.

Pahl-Wostl, C., 1995, *The Dynamic Nature of Ecosystems: Chaos and Order Entwined*, Wiley, Chichester.

Streeter, H.W. and Phelps, E.N., 1925, *A Study of the Pollution and the Natural Purification of the Ohio River*, U.S. Public Health Service, Bulletin No. 146, Columbus, Ohio.

Thoma, J., 1977, *Energy, Entropy and Information*, IISA, Laxenburg, Austria.

Ulanowicz, R., 1991, Formal agency in ecosystem development, in Higashi, M. and Burns, T.P. (Eds.), *Theoretical Studies of Ecosystems: The Network Perspective*, Cambridge University Press, Cambridge, U.K.

Woodwell, G.M., et al., 1967, DDT residues in an east coast estuary: a case of biological concentration of a persistent insecticide, *Science*, 156, 821.

# 3

## System Properties of Ecosystems

S.E. Jørgensen

## CONTENTS

## 3.1 Introduction

System ecology was initiated in the scientific community in the 1950s (see von Bertalanffy, 1950; Odum, 1953), but the real "take off," with society providing a tailwind, came in mid 1960s as a result of the environmental discussion triggered by, among many outstanding environmental books, Rachel Carson's *Silent Spring* (1962). Koestler's book *The Ghost in the Machine* (1967) also contributed to the initial development of ecosystem theory. *Limits to Growth* by Meadows et al. (1972) should also be mentioned in this context as an initiator of environmental discussion, which provoked a further need for new and more holistic ecology.

The interest in ecosystems as complex systems from a holistic viewpoint, meaning that the system is more than the sum of its parts and has characteristic emergent properties, has increased enormously during the last couple of decades due to our concern for the environment. We want to understand the reaction of nature to our impact on the environment and have in this context realized that reductionist methods cannot cope with the environmental problems, mainly because:

1. Unexpected effects, unexpected in time and space, too, may occur due to the many linkages and indirect effects in ecosystems.

2. Ecosystems are extremely complex, which makes them impossible to analyze and know all their details.

Consequently, we have to use another approach, which is called *systems ecology*, where the focus is on the properties of entire ecosystems.

Ecosystems are what is called medium-number systems. They have many components and they are all different. A typical ecosystem would have on the order of $10^{15}$ to $10^{20}$ components. Some of them may belong to the same species or the same type of nonliving components

1-56670-337-9/00/$0.00+$.50
© 2000 by CRC Press LLC

(e.g., suspended clay particles). The number of species in ecosystems may vary considerably, but will usually be on the order of 1000 to 100,000. Individuals belonging to the same species may have some characteristic properties in common, but all individuals are still different from all other individuals. They have, for instance, their own genetic code, which implies that we are dealing with systems with $10^{15}$ to $10^{20}$ components with clearly different properties. This variability is indispensable, because it is the basis for Darwinian selection and, therefore, for evolution. The conclusion is therefore that it would be an impossible task to analyze and know in detail all the components of an ecosystem, and even if we could, the detailed knowledge about all the components may soon be useless, because the biological components are constantly changing as they adapt to the steadily changing conditions. This will be discussed in more detail in the next section.

Complex adaptive systems have several common features (Brown, 1995):

1. They are composed of numerous components of many different kinds.
2. The components interact and react sometimes nonlinearly and on different temporal and spatial scales.
3. The systems organize themselves to produce complex structures and behaviors.
4. The systems maintain thermodynamically unlikely states by the application of their openness.
5. Some form of heritable information allows the systems to respond adaptively to environmental changes.
6. The structure and dynamics of these systems are effectively irreversible, and there is always a legacy of history.
7. They are hierarchically organized.

Consequently, we need a science of ecosystems, which can assist us in our effort to understand their reactions to our impacts on them. This science should be based on ecosystem theories, where we try to understand reactions of ecosystems without knowing all the details, just as we also try to understand human reactions without knowing all the properties of all human cells and their biochemical reactions. System ecology must emphasize the system properties, and ecosystem theories must be based on our possibilities to draw general conclusions from our knowledge about these properties. This scientific approach implies that we will not be able to give a full description of ecosystems covering all their aspects. We can only give a partial description covering some of the aspects. We need, therefore, a pluralistic approach.

## 3.2   Complexity of Ecosystems

The complexity of an ecosystem is formed not only by a high number of interacting components. The complexity will be reviewed in this section by a survey of the different forms of complexity of ecosystems. Nine forms have been identified to capture the complexity of ecosystems (Jørgensen, 1997):

**1. The number of organisms and species on Earth is very high, and they are all different.** We have many different components. We are able to classify all organisms into groups called species. There are many millions of species on Earth (in the order of $10^7$), and there are in the order of $10^{20}$ organisms (the number is, of course, very uncertain). Organisms belonging to the same species have a high degree of similarity, but every organism is nevertheless different from all others, just as each *Homo sapiens* is different from his neighbor.

Complexity certainly increases as the number of components increases, but the number of components is not the only measure of complexity. One mole consists of $6.62 * 10^{23}$ molecules. Yet physicists and chemists are able to predict pressure, temperature, and volume, not in spite of, but because of the large numbers of molecules. The reason is that all the molecules are essentially identical. (There may be a few, but only a few, different types of molecules: oxygen, nitrogen, carbon dioxide, and so on). Interactions of molecules are random, and overall system averages are easily performed. We are able to apply statistical methods to molecules, but not to the much lower number of very different organisms. The individual motions of the more than $10^{23}$ molecules are unknowable, but in thermodynamics it is allowed to average the motions of all the molecules, and that makes predictions possible. When such averaging is impossible, the problem becomes insoluble. The so-called "three bodies" problem (the influences of three bodies on each other's orbits) is already extremely complex.

Ecosystems or the entire ecosphere are, as mentioned above, "medium number systems." They include most systems and are characterized by an intermediate number of *different* components and structured interrelationships among these components. The enormous diversity of organisms may be envisioned as correlated with the immense variety of environments and ecological niches which exist on Earth.

**2. The high number of species gives an extremely high number of possible connections and different relations.** However, a model with many components and a high number of connections is not necessarily more stable than a simple one (May, 1981). It is the governing theory that there is no (simple) relation between stability and diversity. It is possible in nature to find very stable and simple ecosystems, and it is possible to find rather unstable, very diverse ecosystems. May (1972, 1981) claims, that r-selection is associated with a relatively unpredictable environment and simple ecosystems, while K-selection is associated with a relatively predictable environment and a complex biologically crowded community.

It may be concluded that only a few — relative to the number of species — direct connections exist in ecosystems. Indirect effects are, however, very important (Patten, 1991). We may assume, that many direct connections are not needed to render the system stable and too many direct connections may even increase the possibilities for instability, as can be shown by modeling studies.

Reactions of ecosystems to perturbations have been widely discussed in relation to the stability concepts. However, this discussion has, in most cases, not considered the enormous complexity of regulation — and feedback mechanisms.

The stability concept resilience is understood as the ability of the ecosystem to return "to normal" after perturbations. This concept generates more interest in a mathematical discussion of whether equations may be able to return to steady state, but the shortcomings of this concept in a real ecosystem context are clear.

An ecosystem is a soft system that will *never* return to the same point again. It will be able to maintain its functions on the highest possible level, but never with exactly the same biological and chemical components in the same concentrations again. The species composition or the food web may have changed or may not have changed, but at least it will not be the same organisms with exactly the same properties.

In addition, it is unrealistic to consider that the same conditions will occur again. We can observe that an ecosystem has the property of resilience in the sense that ecosystems tend to recover after stress, but a *complete* recovery will never be realized.

The combination of external factors — the impact of the environment on the ecosystem — will never appear again, and even if it did, the internal factors — the components of the ecosystem — have changed meanwhile and cannot react the same way as the previous internal factors did. The concept of resilience is, therefore, not a realistic quantitative concept. If it is used realistically, it is not quantitative, and if it is used quantitatively (e.g., in mathematics), it is not realistic. To a certain extent, resilience covers the ecosystem property of elasticity, but in fact, the ecosystem is more flexible than elastic. It will change to meet the challenge of changing external factors. It will not try to struggle to return to exactly the same situation.

Resistance is another widely applied stability concept. It covers the ability of the ecosystem to resist change when external factors are changed. This concept needs, however, a more rigorous definition and needs to be considered multidimensionally to be able to cope with real ecosystem reactions. An ecosystem will always be changed, when the conditions are changed; the question is, what is changed and how much?

Webster (1979) used models to examine ecosystem reactions to the rate of nutrient recycling. He found that an increase in the amount of recycling relative to input resulted in a decreased margin of stability, faster mean response time, greater resistance (i.e., greater buffer capacity, see the definition below) and less resilience. Increased storage and turnover rates resulted in exactly the same relationships. Increases in both recycling and turnover rates produced opposite results, however, leading to a larger stability margin, faster response time, smaller resistance, and greater resilience.

Gardner and Ashby (1970) examined the influence on stability of connectance (defined as the number of food links in the food web as a fraction of the number of topologically possible links) of large dynamic systems. They suggest that all large, complex, dynamic systems may show the property of being stable up to a critical level of connectances, and then as the connectances increase further, the system suddenly becomes unstable.

O'Neill (1976) suggests that the many regulation mechanisms and spatial heterogeneity should be accounted for, when the stability concepts are applied to explain ecosystem responses. The role of variability in space and time will be touched upon many times and discussed further below as point 6.

These observations explain why it has been very difficult to find a relationship between ecosystem stability in its broadest sense and species diversity. Compare this with Rosenzweig (1971), where almost the same conclusions are drawn.

It is observed that increased phosphorus loading gives decreased diversity, (Ahl and Weiderholm, 1977; Weiderholm, 1980), but very eutrophic lakes *are* very stable. Figure 3.1 gives the result of a statistical analysis from a number of Swedish lakes. The relationship shows a correlation between a number of species and the eutrophication, measured as chlorophyll-a in $\mu$g/l. A similar relationship is obtained between the diversity of the benthic fauna and the phosphorus concentration relative to the depth of the lakes.

Therefore, it seems appropriate to introduce another but similar concept, called *buffer capacity*, $\beta$. It is defined as the inverse change in state variable relative to the change in forcing functions, expressed mathematically as (Jørgensen 1988, 1992,1994, 1997):

$$\beta = 1/(\partial \text{ (State variable)}/\partial \text{ (Forcing function)})$$

*Forcing functions* are the external variables that are driving the system such as discharge of wastewater, precipitation, wind, and so on, while *state variables* are the internal variables that determine the system, such as the concentration of soluble phosphorus, the concentration of zooplankton, and so on.

The concept of buffer capacity has a definition that allows us to quantify in modeling, and it is also applicable to real ecosystems, because it acknowledges that *some* changes will

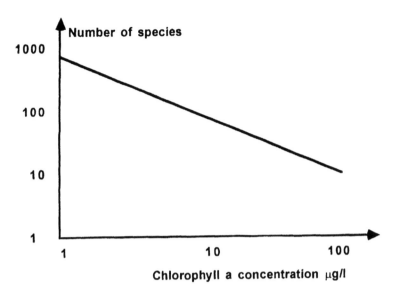

**FIGURE 3.1**
Weiderholm (1980) obtained the relationship shown for a number of Swedish lakes between the number of species and eutrophication, expressed as chlorophyll-a in μg/l.

always take place in the ecosystem in response to changing forcing functions. The question is how large these changes are relative to changes in the conditions (the external variables or forcing functions).

The concept should be considered multidimensionally, so we can consider all combinations of state variables and forcing functions. It implies that even for one type of change there are many buffer capacities corresponding to each of the state variables. Rutledge (1974) defines ecological stability as the ability of the system to resist change in the presence of perturbations. It is a definition very close to that of buffer capacity, but it lacks the multidimensionality of ecological buffer capacity.

The relationship between forcing functions (impacts on the system) and state variables indicating the conditions of the system are rarely linear, and buffer capacities are therefore not constant. So it may be important in environmental management to reveal the relationships between forcing functions and state variables to observe under which conditions buffer capacities are small or large (Figure 3.2).

Model studies (Jørgensen and Meyer, 1977; Jørgensen, 1986) have revealed that in lakes with a high eutrophication level, a high buffer capacity to nutrient inputs is obtained by a relatively small diversity. The low diversity in eutrophic lakes is consistent with the above-mentioned results by Ahl and Weiderholm (1977) and Weiderholm (1980). High nutrient concentrations = large phytoplankton species. The specific surface does not need to be large, because there are plenty of nutrients. The selection or competition is not on the uptake of nutrients but rather on escaping the grazing by zooplankton, and here greater size is an advantage. In other words, the spectrum of selection becomes more narrow, which means reduced diversity. It demonstrates that a high buffer capacity may be accompanied by low diversity.

If a toxic substance is discharged to an ecosystem, the diversity will be reduced. The species most susceptible to the toxic substance will be extinguished, while other species, the survivors, will metabolize, transform, isolate, excrete, etc., the toxic substance and thereby decrease its concentration. We observe a reduced diversity, but simultaneously we maintain a high buffer capacity to input of toxic compounds, which means that only small changes, caused by the toxic substance, will be observed. Model studies of toxic substance

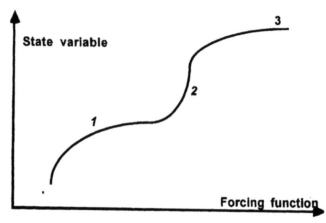

**FIGURE 3.2**

The relation between state variables and forcing functions is shown. At point 1 and 3 the buffer capacity is high; at point 2 it is low.

discharge to a lake (Jørgensen and Meyer, 1977, 1979) demonstrate the same inverse relationship between the buffer capacity to the considered toxic substance and diversity.

Ecosystem stability is a very complex concept (May, 1977), and it seems impossible to find a simple relationship between ecosystem stability and ecosystem properties. Buffer capacity seems to be the most applicable stability concept, because it is based:

1. On an acceptance of the ecological complexity — it is a multidimensional concept
2. On reality, i.e., that an ecosystem will never return to exactly the same situation again

Another consequence of the complexity of ecosystems mentioned above should be considered here. For mathematical ease, the emphasis has been — particularly in population dynamics — on equilibrium models. The dynamic equilibrium conditions (steady state, not thermodynamic equilibrium) may be used as an attractor (in the mathematical sense — the ecological attractor is the thermodynamic equilibrium) for the system, but the equilibrium will never be attained. Before the equilibrium should have been reached, the conditions, determined by the external factors and all ecosystem components, have changed and a new dynamic equilibrium, and thereby a new attractor, is effective. Before this attractor point has been reached, new conditions will again emerge, and so on. A model based on the equilibrium state will therefore give a wrong picture of ecosystem reactions. The reactions are determined by the present values of the state variables and they are different from those in the equilibrium state. We know from many modeling exercises that the model is sensitive to the initial values of the state variables. These initial values are a part of the conditions for further reactions and development. Consequently, the steady-state models may give other results than the dynamic models, and it is therefore recommended to be very careful when drawing conclusions on the basis of equilibrium models. We must accept the complication that ecosystems are dynamic systems and will never attain equilibrium. We need to apply dynamic models as widely as possible, and it can easily be shown that dynamic models give other results than static ones.

**3. The number of feedbacks and regulations is extremely high and makes it possible for the living organisms and populations to survive and reproduce in spite of changes in external conditions.** These regulations form a hierarchy as shown in Table 3.1.

**TABLE 3.1**

The Hierarchy of Regulating Feedback Mechanisms

| Level | Explanation of Regulation Process | Exemplified by Phytoplankton Growth | Timescale |
|---|---|---|---|
| 1 | Rate by concentration in medium | Uptake of phosphorus in accordance with phosphorus concentration | Minutes–hours |
| 2 | Rate by needs | Uptake of phosphorus in accordance with intracellular concentration | Minutes–hours |
| 3 | Rate by other external factors (biochemical adaptation) | Chlorophyll concentration in accordance with previous solar radiation | Days |
| 4 | Adaptation of properties (biological adaptation) | Change of optimal temperature for growth | Days–months |
| 5 | Selection of other species | Shift to better fitted species | Weeks–years |
| 6 | Selection of other food web | Shift to better fitted food web | Months–years |
| 7 | Mutations, new sexual recombinations, and other shifts of genes | Emergence of new species or shifts of species properties | $10-10^5$ years |

Source: Jørgensen, S.E., 1994a, *Fundamentals of Ecological Modelling*, second edition, Elsevier, Amsterdam; Jørgensen, S.E., 1997, *Integration of Ecosystem Theories: A Pattern*, Kluwer Academic Publishers Dordrecht. With permission.

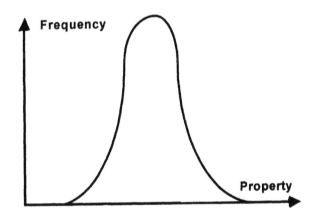

**FIGURE 3.3**
Typical Gaussian frequency distribution of size within the same species.

For example, phytoplankton is able to regulate its chlorophyll concentration according to solar radiation. If more chlorophyll is needed because the radiation is insufficient to guarantee growth, more chlorophyll is produced by the phytoplankton. The digestion efficiency of the food for many animals depends on the abundance of the food. The same species may be of different sizes in different environments, depending on what is most beneficial for survival and growth. If nutrients are scarce, phytoplankton becomes smaller, and vice versa. In this latter case, the change in size is a result of a selection process, which is made possible because of the distribution in size, as illustrated in Figure 3.3.

**4. The feedbacks are constantly changing,** i.e., the adaptation is adaptable in the sense that if a regulation is not sufficient another regulation process higher in the hierarchy of feedbacks (see Table 3.1) will take over. The change in size within the same species is only limited. When this limitation has been reached, other species will take over. It implies that not only the processes and the components, but also the feedbacks can be replaced, if necessary to achieve a better utilization of the available resources. This further supports the application of dynamic models instead of steady-state or equilibrium models, as discussed above. The regulation mechanisms operate at different timescales, which are indicated in the table.

**5. The components and their related processes are organized hierarchically.** This is the well-known hierarchy: genes, cells, organs, organisms, populations, and communities. Processes and regulations will take place on each level in this hierarchy. Each level works as a unit which can be influenced (controlled) from a level higher and lower in the hierarchy.

The hierarchy is established because an increased complexity at a certain level passes through a natural boundary and forms a self-sustainable subunit. The buffer capacity will follow the level of organization, because formation of a unit will imply that a higher level of regulations, adaptations, and feedback mechanisms will occur.

Three different concepts have been used to explain the functioning of ecosystems.

1. *The individualistic or Gleasonian concept* assumes populations to respond independently to an external environment.

2. *The super organism or Clementsian concept* views ecosystems as organisms of a higher order and defines succession as ontogenesis of this super organism (see e.g., on self-organization of ecosystems, Margalef, 1968). Ecosystems and organisms are different, however, in one important aspect. Ecosystems can be dismantled without destroying them; they are just replaced by others, such as agroecosystems, or human settlements, or other successional states. Patten (1991) has pointed out that the indirect effects in ecosystems are significant compared to the direct ones, while in organisms the direct linkages will be most dominant. An ecosystem has more linkages than an organism, but most of them are weaker. It makes the ecosystem less sensitive to the presence of *all* the existing linkages. It does not imply that the linkages in ecosystems are insignificant and do not play a role in ecosystem reactions. The ecological network is of great importance in an ecosystem, but the many and indirect effects give the ecosystem buffer capacities to deal with minor changes in the network. The description of ecosystems as super organisms, therefore, seems insufficient.

3. *The hierarchy theory* (Allen and Starr, 1982) insists that the higher-level systems have emergent properties that are independent of the properties of their lower-level components. This compromise between the two other concepts seems to be consistent with our observations in nature. The hierarchical theory is a very useful tool to understand and describe such complex "medium number" systems as ecosystems (see O' Neill et al., 1986).

During the last decade a debate has arisen on whether "bottom-up" (limitation by resources) or "top-down" (control by predators) effects primarily control the system dynamics. The conclusion of this debate seems to be that both effects control the dynamics of the system. Sometimes the effect of the resources may be most dominant, sometimes the higher levels control the dynamics of the system, and sometimes both effects determine the dynamics of the system.

This conclusion is nicely presented in *Plankton Ecology* by Sommer (1989). In this volume it is clearly demonstrated that the physical conditions (Reynolds, 1989), the resources (Sommer, 1989), the grazers (Sterner, 1989), the predation on zooplankton (Gliwicz and

Pijanowska, 1989), and the parasites (Van Donk, 1989) may all be controlling the phytoplankton community and its succession. The more general conclusion of this excellent contribution to plankton ecology is that ecosystems are very complex and you should be careful in making broad, general simplifications. Each case should be carefully examined before you make your simplifications, which are only valid for the considered case. Everything in an ecosystem is dependent on everything. A profound understanding of ecosystems is only possible if you accept this property of ecosystem complexity. This is the initial condition for modeling, and therefore you can only make simplifications on the basis of a profound knowledge of the particular case, comprising the specific ecosystem and the specific problem in focus.

A network is a result of a hierarchical interpretation of ecosystem relations. The ecosystem and its properties emerge as a result of many simultaneous and parallel focal-level processes, as influenced by even more remote environmental features. It means that the ecosystem itself will be seen by an observer to be factorable into levels. Features of the immediate environment are enclosed in entities of yet larger scale and so on. This implies that the environment of a system includes historical factors as well as immediately cogent ones (Patten, 1982). The history of the ecosystem and its components is therefore important for its reactions and further development. It is one of the main ideas behind Patten's indirect effect that the indirect effect accounts for the "history," while the direct effect only reflects the immediate effect. The importance of the history of the ecosystem and its components emphasizes the need for a dynamic approach and supports the idea that we will never observe the same situation in an ecosystem twice. The history will always be "between" two similar situations. Therefore, as already mentioned above, the equilibrium models may fail in their conclusions, particularly when we want to look into reactions on the system level.

### 6. Ecosystems show a high degree of heterogeneity in space and time.

An ecosystem is a very dynamic system. All its components, and particularly the biological ones, are steadily moving and its properties are steadily modified, which is why an ecosystem will never return to the same situation again.

Furthermore, every point is different from any other point and, therefore, offers different conditions for the various life forms.

This enormous heterogeneity explains why there are so many species on earth. There is, so to speak, an ecological niche for "everyone," and "everyone" may be able to find a niche where he is best suited to utilize the resources.

Ecotones, the transition zones between two ecosystems, offer a particular variability in life conditions, which often results in a particular richness of species diversity. Studies of ecotones have recently drawn much attention from ecologists, because ecotones have pronounced gradients in the external and internal variables, which give a clearer picture of the relation between external and internal variables.

Margalef (1991) claims that ecosystems are anisotropic, meaning that they exhibit properties with different values when measured along axes in different directions. It means that the ecosystem is not homogeneous in relation to properties concerning matter, energy, and information, and that the entire dynamics of the ecosystem works toward increasing the differences.

These variations in time and space make it particularly difficult to model ecosystems and to capture their essential features. However, the hierarchy theory applies these variations to develop a natural hierarchy as a framework for ecosystem descriptions and theory. The strength of the hierarchy theory is that it facilitates the study and modeling of ecosystems.

**7. Ecosystems and their biological components, the species, evolve steadily and in the long-term perspective toward higher complexity.** Darwin's theory describes the competition among species and states that the species that are best suited to the prevailing conditions in the ecosystem will survive. Darwin's theory can, in other words, describe the changes in ecological structure and the species composition.

All species in an ecosystem are confronted with the question: How is it possible to survive or even grow under the prevailing conditions? The prevailing conditions are considered to be *all* factors influencing the species, i.e., all external and internal factors including those originating from other species. This explains coevolution, because any change in the properties of one species will influence the evolution of the other species.

Species are generally more sensitive to stress than functional properties of ecosystems. Schindler (1987) observed in experimental acidification of lakes that functional properties, such as primary production, respiration, and grazing, were relatively insensitive to the effects of a continued exposure to acidification, while early warning signs could be detected at the level of species composition and morphologies. This underlines the importance of development of structural dynamic models able to predict the change in focal properties of the species which would correspond to a shift in species composition.

All natural external and internal factors of ecosystems are dynamic — the conditions are steadily changing, and there are always many species waiting in the wings, ready to take over, if they are better suited to the emerging conditions than the species dominating under the present conditions. There is a wide spectrum of species representing different combinations of properties available for the ecosystem. The question is, which of these species is best able to survive and grow under the present conditions, and which species is best able to survive and grow under the conditions one step further, and two steps further, and so on? The necessity in Monod's sense is given by the prevailing conditions. The species must have genes, or perhaps phenotypes (meaning properties), which match these conditions in order to survive. However, the natural external factors and the genetic pool available for the test may change randomly or by "chance."

New mutations (misprints are produced accidentally) and sexual recombinations (the genes are mixed and shuffled) steadily emerge and give new material to be tested regarding the question: Which species is best suited under the conditions prevailing just now?

These ideas are illustrated in Figure 3.4. The external factors are steadily changing, and some even relatively fast and partly at random, e.g., the meteorological or climatic factors. The species of the system selected among the species available and represented by the genetic pool, which again is slowly but surely changing randomly or by "chance." The selection in Figure 3.4 includes level 4 of Table 3.1. It is a selection of the organisms that possess the properties best suited to the prevailing organisms according to the frequency distribution (see Figure 3.3). *Ecological development* is the changes over time in nature caused by the dynamics of the external factors, giving the system sufficient time for the reactions and including an organization of the network.

*Evolution*, on the other hand, is related to the genetic pool. It is the result of the relation between the dynamics of the external factors and the dynamics of the genetic pool. The external factors steadily change the conditions for survival, and the genetic pool steadily comes up with new solutions to the problem of survival.

Darwin's theory assumes that populations consist of individuals, who:

1. On average produce more offspring than is needed to replace them upon their death. This is the property of high reproduction.

2. Have offspring which resemble their parents more than they resemble randomly chosen individuals in the population. This is the property of inheritance.

3. Vary in heritable traits influencing reproduction and survival (i.e., fitness). This is the property of variation.

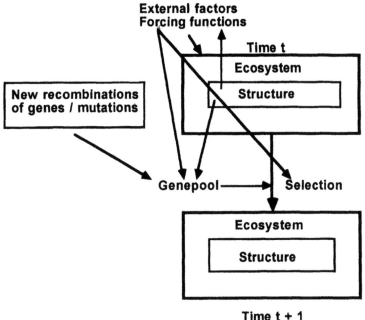

**FIGURE 3.4**

Conceptualization of how the external factors steadily change the species composition. The possible shifts in species composition are determined by the gene pool, which is steadily changed due to mutations and new sexual recombinations of genes. The development is, however, more complex. This is indicated by (1) arrows from "structure" to "external factors" and "selection" to account for the possibility that the species is able to modify its own environment (see below) and thereby its own selection pressure; (2) an arrow from "structure" to "gene pool" to account for the possibilities that the species can to a certain extent change its own gene pool.

All three properties are part of the presentation in Figure 3.4. High reproduction is needed to arrive at a change in the species composition caused by changes in external factors. The variability is represented in the short- and long-term changes in the genetic pool, and the inheritance is needed to see an effect of the fitness test in the long run.

Without the inheritance, every new generation would start from the same point, and it would not be possible to maintain the result of the fitness test. Evolution is able to continue from already-obtained results.

Species are continuously tested against the prevailing conditions (external as well as internal factors), and the better suited they are, the better they are able to maintain and even increase their biomass. The specific rate of population growth may even be used as a measure of the fitness (see Stenseth, 1986), but the property of fitness must of course be inheritable to have any effect on the species composition and the ecological structure of the ecosystem in the long run. Natural selection has been criticized for being a tautology: fitness is measured by survival, and survival of the fittest, therefore, means survival of the survivors. However, the entire Darwinian theory, including the above-mentioned three assumptions, cannot be conceived as a tautology, but may be interpreted as follows: the species offer different solutions to survival under given prevailing conditions, and the species that have the best combinations of properties to match the conditions also have the highest probability of survival and growth. The formulation by Ulanowicz (1986) may also be applied: Those populations are fittest that best enhance the autocatalytic behavior of the matter — energy loops in which they participate.

Man-made changes in external factors, i.e., anthropogenic pollution, have created new problems, because new genes fitted to these changes do not develop overnight, while most

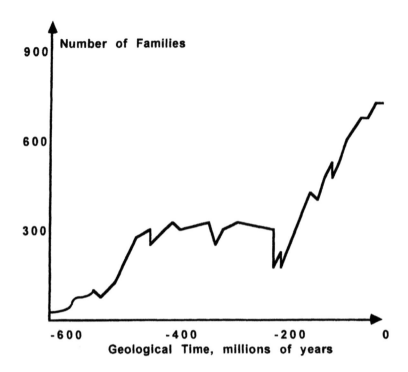

**FIGURE 3.5**
Changes in species diversity over geological time. (Redrawn from Raup, D.M. and Sepkowski, J.J., 1982, Mass extinctions in the marine fossil record, *Science*, 215, 1501. With permission.)

natural changes have occurred many times previously and the genetic pool is therefore prepared to meet the natural changes. The spectrum of genes is able to meet most natural changes, but not all of the man-made changes, because they are new and untested in the ecosystem.

Evolution moves toward increasing complexity in the long run (see Figure 3.5). The fossil records have shown a steady increase of species diversity. There may be destructive forces — for instance, man-made pollution or natural catastrophes — for a shorter time, but the probability that:

1. New and better genes will be developed
2. New ecological niches will be utilized

increases with time. The probability will even (again, excluding the short time perspective) increase faster and faster, as the probability is roughly proportional to the amount of genetic material on which the mutations and new sexual recombinations can be developed.

It is equally important to note that a biological structure is more than an active nonlinear system. In the course of its evolution, the biological structure is continuously changed in such a way that its structural map is itself modified. The overall structure thus becomes a representation of all the information received. Biological structure represents, through its complexity, a synthesis of the information with which it has been in communication (Schoffeniels, 1976). Evolution is perhaps the most discussed topic in biology and ecology, and millions of pages have been written about its ecological implications.

Today the basic facts of evolution are taken for granted, and interest has shifted to more subtle classes of fitness/selection, i.e., toward an understanding of the complexity of the evolutionary processes. One of these classes concerns traits that influence not only the fitness of the individuals possessing them, but the entire population. These traits overtly include social behaviors, such as aggression or cooperation, and activities that, through some modification of the biotic and abiotic environment, feed back to affect the population at large (e.g., pollution and resource depletion).

The coevolution explains the interactive processes among species. It is difficult to observe a coevolution, but it is easy to understand that it plays a major role in the entire evolution process. The coevolution of herbivorous animals and plants is an example. Plants will develop toward a better spreading of seeds and a better defense against herbivorous animals. This will, in the latter case, create a selection of the herbivorous animals that are able to cope with the defense. Therefore, the plants and the herbivorous animals will coevolve.

Coevolution means that the evolution process cannot be described as reductionistic, but that the entire system is evolving. A holistic description of the evolution of the system is needed.

The Darwinian and Neodarwinian theories have been criticized from many sides. It has been questioned whether the selection of the fittest can explain the relatively high rate of evolution. Fitness may here be measured by the ability to grow and reproduce under the prevailing conditions. It implies that the question raised according to the Darwinian theories is: Which species have the properties that give the highest ability for growth and reproduction? The complexity of the evolution processes is often overlooked in this debate. Many interacting processes in the evolution may be able to explain the relatively high rate of evolution.

Six examples are used to illustrate that many processes: (1) interact; (2) accelerate the rate of evolution; and (3) increase the complexity of the evolutionary processes.

1. A mother tiger is an excellent hunter, and she is able to feed many offspring and bring her good "hunting genes" further along the evolution path. Her tiger kittens have a great probability to survive because they get sufficient food. In addition she can teach them her hunting strategy and will have more time to care for them in general because of her successful hunting. So, the kittens not only survive, i.e., the genes survive, but also a better nursing and hunting strategy survives from one tiger generation to the next. We can say in our "computer age" that not only the hardware (the genes) but also the software (the know how) survives.

2. McClintock has observed by working with maize, that genes on chromosomes actually move around or transpose themselves; they even appear to change in relation to environmental stress factors. He proposes the idea that the genetic program is not necessarily fixed in each one. Other geneticists have found what have been dubbed "jumping genes" and to a certain extent confirm this idea. Jumping genes are often named transposons and many workers have labeled them "selfish DNA" (Dawkins, 1989). These discoveries may form the basis for a revolution in biological thinking: the reductionist image of a genetic blueprint may be false.

3. Cairns et al. (1988) showed that when bacteria lacking an enzyme for metabolizing lactose were grown in a lactose medium, some of them underwent a mutation that subsequently enabled them to produce the enzyme. This mutation violated the long-held central dogma of molecular biology, which asserts that information flows only one way in the cell — from genes to RNA to protein and enzyme. Here the information was obviously going in reverse. An enzyme coded for by a particular gene was feeding back to change that gene itself.

4. Symbiosis is generally very well developed in nature. Polycellular organisms are a result of symbiotic relationships among many unicellular organisms, according to Lynn Margulis, as can be recognized from the endosymbiosis in all organisms. It may explain the jumps in the evolution: two or more "properties" are suddenly united and create a symbiotic effect (see Mann, 1991).

5. Fischer and Hinde (1949) describe how the habit of opening milk bottles has spread among blue and great tits. Milk bottles were left on the doorsteps of households and were raided by these songbirds, which open them by tearing off their foil caps. The birds then drink the cream from the top of the bottles. The habit has probably spread through some type of social learning or social enhancement. A novel and learned behavior appears to have modified these birds' environments in ways that have subsequently changed the selection pressures that act back on the birds themselves (Sherry and Galef, 1984). None have shown any genetic response to these altered selection pressures. This example illustrates what Odling-Smee and Patten (1992) call ecological inheritance, which they assert works parallel to the genetic inheritance. The ecological inheritance is a result of species' ability to change their environment and thereby, to a certain extent, modify the selection pressure on themselves.

   Nobody dealing with evolution would deny these possibilities of the species to modify their own environment, but the influence of this ability on the evolution process has most probably been underestimated. Odling-Smee and Patten attempt to emphasize the importance by introducing the concept of "envirotype" as a supplement to genotype and phenotype.

6. A further complication is the so-called morphogenes or D-genes, but it is not possible to go into detail in this text. Further information can be found in Augros and Stanciu (1987), and Dawkins (1982 and 1989).

A total image of evolution will require a holistic approach to account for the many simultaneously interacting processes.

**8. The ecosystems maintain thermodynamically unlikely states.** It is far more complex to describe these states than thermodynamic equilibria. This form of complexity is the focus of the application of thermodynamics in ecology.

**9. The direction and magnitude of any change is affected by preexisting conditions (Brown, 1995).** This implies that structure and dynamics of these systems are effectively irreversible, and there is also a legacy of history. The role of the history of the ecosystem is briefly touched on above, and the importance of the initial conditions is covered by the introduction of chaos theory in ecology (see Jørgensen, 1997).

## 3.3 Quantum Mechanical Uncertainty and Ecosystem Complexity

How can we describe such complex systems as ecosystems in detail? The answer is that it is impossible, if the description must include all details, including all interactions between all the components in the entire hierarchy and all details on feedbacks, adaptations, regulations, and the entire evolution process.

Jørgensen (1988, 1990, 1994b) has introduced the application of the uncertainty principles of quantum mechanics in ecology rooted in the complexity of ecosystems. In nuclear physics the uncertainty is caused by the observer of the incredibly small nuclear particles, while the uncertainty in ecology is caused by the enormous complexity of ecosystems.

For instance, if we take two components and want to know all the relations between them, we would need at least three observations to show whether the relations were linear or nonlinear. Correspondingly, the relations among three components will require 3*3 observations for the shape of the plane. If we have 18 components we would correspondingly need $3^{17}$, or approximately $10^8$, observations. At present this is probably an approximate, practical upper limit to the number of observations which can be invested in one project aimed for one ecosystem. This could be used to formulate a practical uncertainty relation in ecology (see also Jørgensen, 1988):

$$10^5 * \Delta x / \sqrt{3^{n-1}} \approx 1 \tag{3.1}$$

where $\Delta x$ is the relative accuracy of one relation, for instance 0.1 (10% relative), and n is the number of components examined or included in the model.

The 100 million observations could, of course, also be used to give a very exact picture of one relation. Costanza and Sklar (1985) talk about the choice between the two extremes: knowing "everything" about "nothing" or "nothing" about "everything." The first refers to the use of all the observations on one relation to obtain a high accuracy and certainty, while the latter refers to the use of all observations on as many relations as possible in an ecosystem.

How we can obtain a balanced complexity in the description will be further discussed in Chapter 4.

Equation (3.1) formulates a practical uncertainty relation, but, of course, the possibility that the practical number of observations may be increased in the future cannot be excluded. Ever more automatic analytical equipment is steadily emerging. This means that the number of observations that can be invested in one project may be one, two, three, or even several magnitudes larger in one or more decades. However, a theoretical uncertainty relation can be developed. If we go to the limits given by quantum mechanics, the number of variables will still be low, compared to the number of components in an ecosystem.

One of Heisenberg's uncertainty relations is formulated as follows:

$$\Delta s * \Delta p \geq h / 2\pi \tag{3.2}$$

where $\Delta s$ is the uncertainty in determination of the place, and $\Delta p$ is the uncertainty of the momentum. According to this relation, $\Delta x$ of Equation 3.2 should be in the order of $10^{-17}$ if $\Delta s$ and $\Delta p$ are about the same order of magnitude. Another of Heisenberg's uncertainty relations may now be used to give the upper limit of the number of observations:

$$\Delta t * \Delta E \geq h / 2\pi \tag{3.3}$$

where $\Delta t$ is the uncertainty in time and $\Delta E$ in energy.

If we use all the energy that Earth has received during its lifetime of 4.5 billion years we get:

$$173 * 10^{15} * 4.5 * 10^9 * 365.3 * 24 * 3600 = 2.5 * 10^{34} \text{ J}, \tag{3.4}$$

where $173 * 10^{15}$ W is the energy flow of solar radiation. $\Delta t$ would, therefore, be in the order of $10^{-69}$ sec. Consequently, an observation will take at least $10^{-69}$ sec, even if we use all the energy that has been available on Earth as $\Delta E$, which must be considered the most extreme case. The hypothetical number of observations possible during the lifetime of the Earth would therefore be:

$$4.5 * 10^9 * 365.3 * 3600 / 10^{-69} \approx \text{of } 10^{85} \qquad (3.5)$$

This implies that we can replace $10^5$ in Equation 3.1 with $10^{60}$ since

$$10^{-17} / \sqrt{10^{85}} \approx 10^{-60}$$

If we use $\Delta x = 1$ in Equation 3.1, we get:

$$\sqrt{3^{n-1}} \leq 10^{60} \qquad (3.6)$$

or

$$n \leq 253.$$

From these very theoretical considerations, we can clearly conclude that we shall never be able to get a sufficient number of observations to describe even one ecosystem in all detail. These results are completely in harmony with Niels Bohr's complementarity theory. He expressed it as follows: "It is not possible to make one unambiguous picture (model) of reality, as uncertainty limits our knowledge." The uncertainty in nuclear physics is caused by the inevitable influence of the observer on the nuclear particles; in ecology, it is caused by the enormous complexity and variability.

Quantum theory may have an even wider application in ecological theories. Schrödinger (1944) suggests that the "jump-like changes" you observe in the properties of species are comparable to the jump-like changes in energy by nuclear particles. Schrödinger was inclined to call De Vries' mutation theory (published in 1902), the quantum theory of biology, because the mutations are due to quantum jumps in the gene molecule.

Patten (1982) defines an elementary "particle" of the environment, called an *environ* (previously he used the word *holon*) as a unit which is able to transfer an input to an output. Patten suggests that a characteristic feature of ecosystems is the connectances. Input signals go into the ecosystem components, and they are translated into output signals. Such a "translator unit" is an environmental quantum, according to Patten. The concept is borrowed from Koestler (1967), who introduced the word *holon* to designate the unit on a hierarchic tree. The term comes from Greek *holos* = whole, with the suffix *on*, as in proton, electron, and neutron, suggesting a particle or part.

Stonier (1990) introduces the term *infon* for the elementary particle of information. He envisages an infon as a photon whose wavelength has been stretched to infinity. At velocities other than c, its wavelength *appears* infinite, its frequency zero. Once an infon is accelerated to the speed of light, it crosses a threshold, which allows it to be perceived as having energy. When that happens, the energy becomes a function of its frequency. Conversely, at velocities other than c, the particle exhibits neither energy nor momentum, yet it could retain at least two information properties: its speed and its direction. In other words, at velocities other than c, a quantum of energy becomes converted into a quantum of information — an infon.

## 3.4  Complementarity in Ecology

The use of maps in geography is a good parallel to the situation in ecology. Just as we have road maps, airplane maps, geological maps, maps in different scales for different purposes, we have in ecology many models of the same ecosystems and we need them all, if we want to get a comprehensive, pluralistic view of ecosystems. Furthermore, a map can never give a complete picture. We can always do the scale larger and larger and include more details, but we cannot get all the details. Even if we could, the picture would be invalid a few seconds later because we try to map too many dynamic details at the same time. An ecosystem also consists of too many dynamic components to enable us to model all the components simultaneously and even if we could, the model would be invalid a few seconds later, because the dynamics of the system have changed the "picture."

In nuclear physics, we need to use many different pictures of the same phenomena to be able to describe our observations. We say that we need a pluralistic view to cover our observations completely. Our observations of light, for instance, require that we consider light as waves as well as particles. The situation in ecology is similar. Because of the immense complexity, we need a pluralistic view to cover a description of the ecosystems according to our observations. We need many models covering different viewpoints. Descriptions of an ecosystem as a network by use of population dynamics, as an evolutionary system by use of a series of different models, and so on, are all valid images of the ecosystem, and they all contribute to a better understanding of how the system works. One description is not necessarily better than another, but they all contribute to a more comprehensive image of the ecosystem. They are complementary. All the possible descriptions of an ecosystem cover different aspects of the ecosystem, and together they give the best description we can give of the system, but due to the enormous complexity of the system, all these descriptions will not together give a complete, accurate picture. All the descriptions can still be more complete or more accurate, and additional descriptions not known today may improve our understanding of the system in the future. We will probably steadily improve our understanding of the system but never attain a complete accurate description due to its complexity. And even if we could, this understanding will be invalidated in a short time, because the system is dynamic and constantly changing. It is consistent with Gödel's Theorem from 1931 (see Gödel, 1989) that the infinite truth can never be condensed in a finite theory. There are limits to our insight, or we cannot produce a map of the world with all possible details, because that would be the world itself.

Furthermore, ecosystems must be considered as irreducible systems (Wolfram, 1984a, b) in the sense that it is not possible to make observations and then reduce the observations to more or less simple laws of nature, as is true in mechanics for instance. Too many interacting components force us to consider ecosystems as irreducible systems. The entire ecological network plays a role for all the processes in an ecosystem. If we isolate a few components and their interacting processes by a laboratory or *in situ* experiment, we will exclude the indirect effects of the components interacting through the entire network. As the indirect effects often are more dominant than the direct ones, our experiment will not be able to reveal the results of the relations as they are observed in nature. The same problem is found today in nuclear physics, where the picture of the atoms is now "a chaos" of many interacting elementary particles. Assumptions on how the particles are interacting are formulated as models, which are tested toward observations. We draw upon exactly the same solution to the problem of complexity in ecology. It is necessary to use what is called experimental mathematics or modeling to cope with such irreducible systems. Today, this is the tool in nuclear physics, and the same tool is increasingly used in ecology (Jørgensen, 1994a).

## 3.5 Holism vs. Reductionism

Holism and reductionism are two different approaches to revealing the secrecies of nature.

**Holism** attempts to reveal the properties of complex systems such as ecosystems by studying the systems as a whole. According to this approach, the system properties cannot be found by studying the components separately because of the high complexity and the presence of emergent properties. Therefore, although it is far more difficult, it is required that the study be on the system level. This does *not* imply that a good knowledge of the components and their properties is redundant. The more we know about the system on all levels, the better we are able to extract the system properties. But it *does* imply, that a study of the components of ecosystems will never be sufficient, because such a study will never reveal the system properties. The components of ecosystems are coevolutionary, coordinated to such an extent that ecosystems work as indivisible unities.

**Reductionism** attempts to reveal the properties of nature by separating the components from their wholeness to simplify the study and to facilitate the interpretation of the scientific results. This scientific method has been the core approach since Newton. It is, of course, always very useful to find governing relationships in nature, for instance, primary production vs. radiation intensity, mortality vs. concentration of a toxic substance, etc. However, the method has obvious shortcomings, when the functions of entire ecosystems are to be revealed. A human being cannot be described on the basis of the properties of all the cells of the body. The function of a church cannot be found through studies of the bricks, the columns, etc. There are numerous examples of the additional need for holistic approaches.

It is far easier to take a complex system apart than it is to reassemble the parts and restore the important functions. Similarly, no science has succeeded in understanding the structure and dynamics of a complex system from a reductionistic approach alone. During the last few decades, population ecologists have tried to describe and predict the fluctuations in the local abundance of species, and community ecologists have tried to describe and predict changes in the composition of locally coexisting species. Progress has been slow and generality has been limited, not because the cause of past changes in abundance or species composition are inherently unknowable or because prediction of the future trajectory is impossible, but rather because even small differences can be amplified by nonlinear processes to produce divergent outcomes.

The conclusion from these considerations is clear: we need both approaches, but because it is much easier to apply the reductionistic method, analytical work has been overwhelming synthetic work in science, particularly during the period from 1945 to 1975. The last 25 years of ecological research have shown with increasing clarity that the need for the holistic approach is urgent. Many ecologists feel that a holistic ecosystem theory is a necessary basis for a more comprehensive understanding of the ecosphere and the ecosystems and for a solution to all threatening global problems.

The need for a more holistic approach increases with the complexity, integration, number of interactions, feedbacks and regulation mechanisms. A mechanical system (a watch, for instance) is divisible, while an ecosystem is indivisible, because of the well-developed interdependence. The ecosystem has developed this interdependence over billions of years. All species have evolved step by step by selection of a set of properties which are evolved from all the prevailing conditions (i.e., all external factors and all other species). All species are influenced to a certain extent by all other biological and nonbiological components of the ecosystem. All species are therefore confronted with the question: "Which of the possible *combinations* of properties will give the best chance for survival and growth, considering all possible factors, i.e., all forcing functions and all other components of the ecosystem?" That combination will be selected and that will be the combination that gives the best benefit in the long run to the entire system, as all components try to optimize the answer to the same question. This game has continued for billions of years. A steady refinement of the properties has taken place, and it has been possible through this evolution to consider ever more factors, which means that the species have become increasingly integrated with the system and ever more interactions have developed.

Patten (1991) expresses numerically the direct and the indirect effect. The direct effect between two components in an ecosystem is the effect of the direct link between the two components. The link between phytoplankton and zooplankton, for instance, is the grazing process. The indirect effect is the effect caused by all relationships between two components except the direct one. The grazing of zooplankton also has a beneficial effect on phytoplankton, because the grazing will accelerate the turnover rate of the nutrients. It is difficult mathematically to consider the total indirect effect to be able to compare it with the direct effect. It can, however, be revealed in this context that Patten has found that the indirect effect often may be larger than the direct one. It implies that a separation of two related components in an ecosystem for examination of the link between them will not be able to account for a significant part of the total effect of the relationship. The conclusion from Patten's work is clearly that it is *not* possible to study an ecosystem on the system level, taking all interrelations into account by studying the direct links only. An ecosystem is more than the sum of its parts.

Lovelock (1979) has taken a full step in the holistic direction. He considers our planet to be one cooperative unit. Its properties cannot be understood, in his opinion, without an assumption of a coordinated coevolution of the approximately 10 million species on Earth.

Lovelock (1988) was struck by the unusual composition of the atmosphere. How could methane and oxygen be present simultaneously? Under normal circumstances these two gases would react readily to produce carbon dioxide and water. Looking further, he found that the concentration of carbon dioxide was much smaller on Earth than if the atmospheric gases had been allowed to go into equilibrium. The same is true for the salt concentration in the sea. Lovelock concluded that the planet's persistent state of disequilibrium was clear proof of life activities and that the regulations of the composition of the spheres on Earth have coevolved over time. Particularly the cycling of essential elements has been regulated to the benefit of life on Earth. Lovelock believes that innumerable regulating biomechanisms are responsible for the homeostasis or steady-state far-from-equilibrium of the planet. Three examples will be mentioned here to illustrate this challenging idea further.

Ocean plankton emits a sulfurous gas into the atmosphere. A physical–chemical reaction transforms the gas into aerosols on which water vapor condenses, setting the stage for cloud formation. The clouds then reflect a part of the sunlight back into space. If Earth becomes too cool, the number of plankton is cut back by the chill. The cloud formation is thereby reduced, and the temperature rises. The plankton operates like a thermostat to keep the Earth's temperature within a certain range.

The silica concentration of the sea is controlled by diatoms. Less than 1% of the silica transported to the sea is maintained at the surface. Diatoms take up the silica, and when they die they settle and remove the silica from the water to the sediment. The composition of the sea is maintained far from the equilibrium known from salt lakes without life, due to the presence of diatoms. Life — the diatoms — dictates that life conditions are maintained in the sea.

Sulfur is transported from the lithosphere into the sea causing an imbalance. If there were no regulations, the sulfur concentration of the sea would be too high, and sulfur would be lacking in the lithosphere as an essential element. However, many aquatic organisms are able to get rid of undesired elements by methylation processes. Methyl compounds of mercury, arsenic, and sulfur are very volatile, which implies that these elements are transported from the hydrosphere to the atmosphere by methylation processes. *Polysiphonia fastigiata*, a marine alga, is capable of producing a huge amount of dimethylsulfide (Lovelock, 1979). This biological methylation of sulfur seems able to explain that the delicate balance of essential elements between the spheres is maintained.

The Gaia hypothesis presumes that the components of the ecosphere and therefore also of the ecosystems cooperate more than they compete, when we contemplate the effects from a system's viewpoint. This point is illustrated by an example which shows how symbiosis can develop and lead to new species. The example is described by Barlow (1991) and the event was witnessed and described by Kwang Jeon. Kwang Jeon had been raising amoebas for years, when he received a new batch for his experiments. The new batch disseminated a severe illness, and the amoebas refused to eat and failed to reproduce. Many amoebas and the few that grew and divided did so reluctantly. A close inspection revealed that about 100,000 rod-shaped bacteria, brought in by the new amoebas, were present in each amoeba. The surviving bacterized amoebas were fragile. They were easily killed by antibiotics and were oversensitive to heat and starvation. For five years, Jeon nurtured the infected amoebas back to health by continuously selecting the tougher ones. Even though they were still infected, they started to divide again at the normal rate. They had not got rid of their bacteria, but they had adapted and were cured of their disease. Each recovered amoeba contained about 40,000 bacteria, and they have adjusted their destructive tendencies in order to live and survive inside other living cells.

From friends Jeon reclaimed some of the amoebas that he had sent off before the epidemic. With a hooked glass needle he removed the nuclei from the infected and uninfected organisms and exchanged them. The infected amoebas with new nuclei survived, while the uninfected amoebas supplied with nuclei from cells that had been infected for years struggled for about four days and then died. The nuclei were unable to cope with an uninfected cell. To test this hypothesis Jeon injected uninfected cells with nuclei from infected amoebas with a few bacteria, just before they died. The bacteria rapidly increased to 40,000 per cell, and the amoebas returned to health. Obviously, a symbiosis had developed.

The amoeba experiment shows that cooperation is an extremely important element in evolution. An ultimate cooperation of all components in the ecosystems leads inevitably to a Gaia perception of ecosystems and the entire ecosphere.

It is interesting that Axelrod (1984) demonstrates through the use of game theory that cooperation is a beneficial long-term strategy. The game anticipates a trade situation between you and a dealer. At mutual cooperation, both parties earn two points, while at mutual defection, both earn zero points. Cooperating while the other part defects stings: you get minus one point, while the "rat" gets something for nothing and earns four points. Should you happen to be a rat, while the dealer is cooperative, you get four points and the dealer loses one. Which strategy should you follow to gain most? Two computer tournaments have given the result that the following "tit-for-tat" strategy seems to be winning: start with a cooperative choice and then do what the other player did in the previous move. In other words, be open for cooperation unless the dealer is not, but only defect one time after the dealer has defected.

It may be possible to conclude that the acceptance of the Gaia hypothesis does not involve that mysterious, unknown, global forces are needed to be able to explain these observations of homeostasis. It seems to be possible to explain the hypothesis by an evolution based upon five factors:

1. Selection (steadily ongoing test of which properties give the highest chance of survival and growth) from a range of properties, offered by the existing species.

2. Interactions of randomness (new mutations and sexual genetic recombinations are steadily produced) and necessity, i.e., to have the right properties for survival under the prevailing conditions, resulting from all external factors and all other components of the ecosystem.

3. A very long time has been available for this ongoing "trial and error" process, which has developed the ecosphere step-wise toward the present, ingenious complexity, where all components have unique and integrated properties.

4. The ability of the biological components to maintain the results already achieved (by means of genes) and to build upon these results in the effort to develop further.

5. As the complexity of the ecosystems and thereby of the entire ecosphere develops, the indirect effect becomes more and more important. It implies that the selection based upon the "effects" on the considered component will be determined by the entire ecosystem and that this selection process will assure that all components of the ecosystem will evolve toward being better and better suited to the entire ecosystem. It means that the system will evolve toward working more and more as a whole — as an integrated system — and that the selection will be more and more beneficial for the entire system.

---

# References

Ahl, T. and Weiderholm, T., 1977, Svenska vattenkvalitetskriterier: Eurofierande ämnen, Svenska Naturritens Kaps Akademi, Sweden.

Allen, T.F.H. and Starr, T.B., 1982, *Hierarchy: Perspectives for Ecological Complexity*, University of Chicago Press, Chicago, IL.

Augros, R. and Stanciu, G., 1987, *The New Biology: Discovering the Wisdom of Nature*, Shambhala, Boston, MA.

Axelrod, R., 1984, *The Evolution of Cooperation*. Basic Books, New York.

Barlow, C. (Editor), 1991, *From Gaia to Selfish Genes: Selected Writings in the Life Sciences*, MIT Press, Cambridge, MA.

Brown, H., 1995, *Macroecology*, University of Chicago Press, Chicago, IL.

Cairns, J., Overbaugh, J., and Miller, S., 1988, The origin of mutants, *Nature*, 355, 142.

Carson, R., 1962, *Silent Spring*, New American Library, New York.

Constanza, R. and Sklar, F.H., 1985, Articulation, accuracy and effectiveness of mathematical models: a review of freshwater wetland applications, *Ecol. Modelling*, 27, 45.

Dawkins, R.D., 1982, *The Extended Phenotype*, Freeman, Oxford.

Dawkins, R.D., 1989, *The Selfish Gene*, second edition, Oxford University Press, Oxford.

Fisher, J. and Hinde, R.A., 1949, The opening of milk bottles by birds, *Br. Birds*, 42, 347.

Gardner, M.R. and Ashby, W.R., 1970, Connectance of large dynamical (cybernetic) systems: critical values for stability, *Nature*, 288, 784.

Gliwicz Z.M. and Pijanowska, J., 1989, The role of predation in zooplankton succession, in U. Sommer (editor), *Plankton Ecology: Succession in Plankton Communities*, Springer-Verlag, Berlin.

Gödel, K., 1989, *Collected Works. Vol. 1*, S. Feferman et al. (editors), Oxford University Press, New York.

Jørgensen, S.E., 1986, Structural dynamic model, *Ecol. Modelling*, 31, 1.

Jørgensen, S.E., 1988, Use of models as an experimental tool to show that structural changes are accompanied by increased exergy, *Ecol. Modelling*, 41, 117.

Jørgensen, S.E., 1990, Ecosystem theory, ecological buffer capacity, uncertainty and complexity, *Ecol. Modelling*, 52, 125.

Jørgensen, S.E., 1994a, *Fundamentals of Ecological Modelling*, second edition, Elsevier, Amsterdam.

Jørgensen, S.E., 1994b, Models as instruments for combination of ecological theory and environmental practice, *Ecol. Modelling*, 75/76, 5.

Jørgensen, S.E., 1997, *Integration of Ecosystem Theories: A Pattern*, Kluwer Academic Publishers Dordrecht.

Jørgensen, S.E. and Meyer, H.F., 1977, Ecological buffer capacity, *Ecol. Modelling*, 3, 39.

Jørgensen, S.E. and Mejer, H.F., 1979, A holistic approach to ecological modelling, *Ecol. Modelling*, 7, 169.

Koestler, A., 1967, *The Ghost in the Machine*, Macmillan, New York.

Lovelock, J.E., 1979, *Gaia, a New Look at Natural History*, Oxford University Press, Oxford.

Lovelock, J.E., 1988, *The Ages of Gaia*. Oxford University Press, Oxford.

Mann, C., 1991, Lynn Margulis: Science's unruly earth mother, *Science*, 252, 378.

Margalef, R., 1991, Networks in ecology, in Higashi, M. and Burns, T.P., Eds., *Theoretical Studies of Ecosystems: The Network Perspective*. Cambridge University Press, Cambridge, 41.

May, R.M., 1972, Will a large complex system be stable? *Nature*, 238, 413.

May, R.M., 1977, *Stability and Complexity in Model Ecosystems*, third edition, Princeton University Press, Princeton, NJ.

May, R.M. (editor), 1981, *Theoretical Ecology: Principles and Applications*, (second edition), Blackwell Scientific, Oxford.

Meadows, D.H., Meadows, D.L., Randers, J., and Behrens, W.W., 1972, *The Limits to Growth: A Report for the Club of Rome's Project on the Predicament of Mankind*. Earth Island, London.

Odum, E.P., 1953, *Fundamentals of Ecology*. Saunders, Philadelphia, PA.

Odling-Smee, S. and Patten, B.C., 1992, personal communication.

O'Neill, R.V., 1976, Ecosystem persistence and heterotrophic regulation, *Ecology*, 57, 1244.

O'Neill, R.V., DeAngelis, D.L., Waide, J.B., and Allen, T.F.H., 1986, *A Hierarchical Concept of Ecosystems*, Princeton University Press, Princeton, NJ.

Patten, B.C., 1982, Indirect causality in ecosystem: Its significance for environmental protection, in Mason, W.T. and Iker, S. Eds., *Research on Fish and Wildlife Habitat*, (Commemorative monograph honoring the first decade of the U.S. Environmental Protection Agency, EPA-600/8-82-022, Office of Research and Development, U.S. Environmental Protection Agency, Washington, D.C.

Patten, B.C., 1991, Network ecology: Indirect determination of the life-environment relationship in ecosystems, in Higashi, M. and Burns, T.P., Eds., *Theoretical Studies of Ecosystems: The Network Perspective*, Cambridge University Press, 288.

Raup, D.M. and Sepkowski, J.J., 1982, Mass extinctions in the marine fossil record, *Science*, 215, 1501.

Reynolds, C.S., 1989, Physical determinants of phytoplankton succession, in Sommer, U., Ed. *Plankton Ecology: Succession in Plankton Communities*, Springer-Verlag, Berlin, 9.

Rosenzweig, M.L., 1971, Paradox of enrichment: Destabilization of exploitation ecosystems in ecological time, *Science*, 171, 385.

Rutledge, R.W., 1974, *Ecological Stability: A Systems Theory Viewpoint* Thesis, Oklahoma State University, Oklahoma.

Schindler, D.W., 1987, Effects of acid rain on freshwater ecosystems, *Science*, 239, 149.

Schoffeniels, E., 1976, *Anti-Chance*, Pergamon Press, New York.

Schrödinger, E., 1944, *What is Life?* Cambridge University Press, Cambridge.

Sherry, D.F. and Galef, B.G., 1984, Cultural transmission without imitation: milk bottle opening by birds, *Animal Behav.*, 32, 937.

Sommer, U., 1989, Toward a Darwinian ecology of plankton, in Sommer, U., Ed., *Plankton Ecology: Succession in Plankton Communities*, Springer-Verlag, Berlin, 1.

Stenseth, N.C., 1986, Darwinian evolution in ecosystems: a survey of some ideas and difficulties together with some possible solutions, in Casti, J.L. and Karlqvist, A., Eds., *Complexity, Language, and Life: Mathematical Approaches*, Springer-Verlag, Berlin, 105.

Sterner, R.W., 1989, The role of grazers in phytoplankton succession, in Sommer, U., Ed., *Plankton Ecology: Succession in Plankton Communities*, Springer-Verlag, Berlin, 107.

Stonier, T., 1990, *Information and the Internal Structure of the Universe*, Springer-Verlag, London.

Ulanowicz, R.E., 1986, *Growth and Development, Ecosystems Phenomenology*, Springer-Verlag, New York.

Van Donk, E., 1989, The role of fungal parasites in phytoplankton succession, in Sommer, U., Ed., *Plankton Ecology: Succession in Plankton Communities*, Springer-Verlag, Berlin, 171.

von Bertalanffy, L., 1950, An outline of general systems theory, *Br. J. Phil. Sci.*, 1, 134.

Webster, J.R., 1979, Hierarchical organization of ecosystems, in Halfon, E., Ed., *Theoretical Systems Ecology*, Academic Press, New York, 119.

Weiderholm, T., 1980, Use of benthos in lake monitoring, *J. Water Pollut. Control Fed.*, 52, 537.

Wolfram, S., 1984a, Cellular automata as models of complexity, *Nature*, 311, 419.

Wolfram, S., 1984b, Computer software in science and mathematics, *Sci. Am.*, 251, 140.

# 4

## Ecosystem Health

S.E. Jørgensen and S.N. Nielsen

## CONTENTS

## 4.1  Ecosystem Health and Integrity

More and more environmental managers want to include ecological considerations in their management strategy, so they have asked the following question of ecologists and system ecologists: How can we express and measure that an ecosystem is ecologically sound? The physician attempts to express the health condition of his patient by using indicators, such as blood pressure, temperature, kidney function, etc. The environmental manager is likewise searching for ecological indicators that can assess the condition of ecosystems. Because an ecosystem is a very complex system, it is not surprising that it is not an easy task to find good ecological indicators to give the appropriate information on ecosystem health, although many ecologists and system ecologists have been and are working with this problem. Rapport (1995) uses the phrases "to take nature's pulse," "the problem of detecting diseases in nature," and "clinical ecology" to stress the parallelism to human pathology.

The concept definition of ecosystem health may be summarized, according to Constanza (1992), as follows:

1. Homeostasis
2. Absence of disease
3. Diversity or complexity
4. Stability or resilience
5. Vigor or scope for growth
6. Balance between system components.

1-56670-337-9/00/$0 00+$.50
© 2000 by CRC Press LLC

He emphasizes, that it is necessary to consider all or at least most of the definitions simultaneously. Consequently, he proposes an overall system health index, HI = V*O*R, where V is system vigor, O is the system organization index, and R is the resilience index. With this proposal, Constanza touches on the most crucial ecosystem properties to cover ecosystem health.

Kay (1991) uses the term "ecosystem integrity" to refer to its ability to maintain its organization. Measures of integrity should therefore reflect the two aspects of the organizational state of an ecosystem: the functional and structural aspects. Function refers to the overall activities of the ecosystem. Structure refers to the interconnection between the components of the system. Measures of function would indicate the amount of energy being captured by the system. This could be covered by measuring the exergy captured by the system. Measures of structure would indicate the way in which energy is moving through the system. The exergy stored in the ecosystem could be a reasonable indicator of the structure.

Kay (1991) presents the fundamental hypothesis, that ecosystems will organize themselves to maximize the degradation of the available work (exergy) in incoming energy. A corollary is that material flow cycles will tend to close. This is necessary to ensure a continued supply of material for the energy-degrading processes. Maximum degradation of exergy is a consequence of the development of ecosystems from the early to the mature state, but as ecosystems cannot degrade more energy than corresponds to the incoming solar radiation, maximum degradation may not be an appropriate goal function for the *mature* ecosystems. It should, however, be underlined here that the use of satellite images to indicate where an ecosystem may be found on a scale from an early to a mature system, is a very useful method to assess ecosystem integrity. These concepts have been applied by Akbari (1995) to analyze a nonagricultural and an agricultural ecosystem. He found that the latter system, representing an ecosystem at an early stage, has higher surface canopy-air temperature (less exergy is captured) and less biomass (less stored exergy) than the nonagricultural ecosystem, which represents the more mature ecosystem.

Kay (1991) looks into the organizational reactions of ecosystems in the face of changing environmental conditions. If an ecosystem is able to maintain its organization in spite of changing environmental conditions, the ecosystem is said to have integrity. If an ecosystem is unable to maintain its organization, then it has lost its integrity. Integrity has therefore to do with the ability of the system to attain and maintain its optimum operating point. Kay distinguishes three possibilities, when the environmental changes cause the ecosystem to move from its original optimum operating point:

1. The new optimum operating point is on the original thermodynamic branch. It means that the lower levels (1 to 4) of the regulation mechanisms presented in Table 3.1 can cope with the environmental changes.

2. The new optimum operating point is on a bifurcation from the original branch. It will often correspond to regulation mechanisms level 5 in Table 3.1. An example is the changes in species composition of phytoplankton as a response to a higher or lower input of nutrients to a lake.

3. The new optimum operating point is on a different thermodynamic branch, and the system undergoes a catastrophic reorganization to reach it. An example is the elimination of top-carnivorous fish caused by hypereutrophication or by acid rain.

Kay (1991) illustrates the three cases as shown in Figure 4.1. In all three cases the ecosystems attempt under different constraints to utilize as much of the incoming solar radiation

**TABLE 4.1**

Ecosystem Indicators for the Crystal River March Gut Ecosystem.
Unit: mg/m²/day

| Indicator | Control | Stressed Ecosystem |
|---|---|---|
| Biomass | 1,157,000 | 755,000 |
| **Total** | | |
| Imports | 7,400 | 6,000 |
| throughput | 22,800 | 18,000 |
| Production | 3,300 | 2,600 |
| Exports | 950 | 870 |
| Respiration | 6,400 | 5,100 |
| **Living matter** | | |
| Production | 400 | 330 |
| Exports | 320 | 250 |
| Respiration | 3,600 | 3,100 |
| To detritus | 5,700 | 4,300 |
| **Detritus** | | |
| Import | 0 | 0 |
| Production | 2,900 | 2,200 |
| Exports | 640 | 620 |
| Respiration | 2,800 | 2,100 |
| **Food web** | | |
| Cycles | 142 | 69 |
| Nexuses | 49 | 36 |
| Cycling matter in % of through put | 10,2% | 9,3% |

Source: Kay, J.J. and Schneider, E.D., 1990, On the applicability of non-equilibrium thermodynamics to living systems [internal paper], University of Waterloo, Ontario, Canada. With permission.

as possible for maintenance of (1) as much structure, (2) as large a network, and (3) as many highly specialized resource niches as possible.

Table 4.1 compares a set of ecological indicators (Kay, 1991) for a stressed and a normal ecosystem. The stressed system has flows that are about 20% less (except export that drops by about 8%, but measured relative to the import, it even increases) and the biomass drops about 35%. Wulff and Ulanowicz (1989) compared the Baltic Sea and the Chesapeake Bay using flow analysis techniques. The Baltic Sea has a lower species diversity. Its primary production is three times smaller than that of Chesapeake Bay, and its total system throughput is only 20% of Chesapeake Bay's. These traditional measures would indicate that the Baltic has lower ecological integrity, but Wulff and Ulanowicz found that the Baltic Sea is trophically more efficient and possesses a more highly structured array of recycling loops. Species in the Baltic Sea are higher situated trophically, which is indicated by the number of cycles of each cycle length for the two ecosystems. Chesapeake Bay has more cycles of the length of 3 than the Baltic Sea, while the Baltic Sea has more cycles of the length of 4 and 5 and even has cycles of the length of 6, which cannot be found in Chesapeake Bay.

From these examples and others presented in this chapter, it can be concluded that assessment of ecosystem health is not an easy task. It may be necessary to apply several indicators simultaneously to get a sufficient image of the health or integrity of an ecosystem, which is not surprising. A doctor of medicine uses several indicators to assess the health of his patient. An ecosystem, which is even more complex than the human body, needs a wide spectrum of indicators. It is completely consistent with the complementarity theory and the red thread of system thinking. Various researchers in systems ecology have proposed different indicators that cover different aspects of ecosystem health. Only

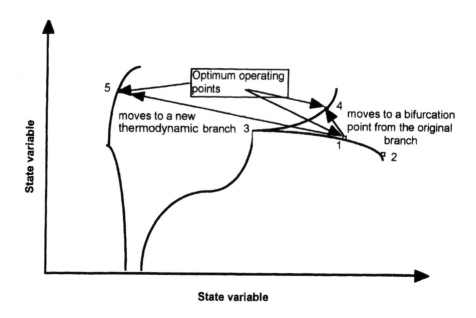

**State variable**

**FIGURE 4.1**

The environmental change may drive the ecosystem from its original optimum operating point to a new point, e.g., from 1 to 2. The environmental conditions may also move the system away from the original operating point, 1, through a bifurcation point, 3, and to a new operating point, 4, via a new path. The environmental changes may also drive the original optimum operating point, 1, to a new thermodynamic branch to the operating point, 5. (Reproduced from three figures in Kay, J.J., 1991, A non-equilibrium thermodynamic framework for discussing ecosystem integrity, *Environ. Manage.*, 15, 483. With permission.)

a multiview description will be able to capture all the features needed to give a fully informative assessment of the conditions of an ecosystem.

## 4.2   Application of Exergy and Buffer Capacities as Ecological Indicators

Exergy expresses, in accordance with Section 2.2, the biomass of the system and the genetic information because this biomass is carrying. Only a relative exergy index, however, can be calculated, because it is impossible to determine the concentrations of all components in an ecosystem. It will measure the relative, approximate distance from thermodynamic equilibrium, but it is obviously only based on the ecological components included in the calculation. It includes the exergy embedded in the ability to make ordering processes, which are carried out by the information stored in the genes. Exergy also expresses the energy needed to decompose the system to inorganic matter (Svirezhev, 1992), and the work the system can perform by a proper use of these decomposition processes.

The relative exergy index may be used as a measure for ecosystem health and will at least partly cover (1), (2), partly (3), (4), and (5) of the six points given in the definitions by Constanza (1992):

1. The homeostasis is embodied in the information of the organisms on how they will meet certain changes by feedback reactions. This information is determined mainly by the genes, which are considered in the calculation of exergy.

2. Absence of disease is reflected in the biomass, as an ecosystem disease sooner or later will be accompanied by a drop in biomass. Because the higher organisms have more genes, exergy will be particularly sensitive to a drop in biomass of these organisms, which is considered an advantage of the use of exergy as an ecological indicator.

3. Living matter has higher complexity compared with the same elements in organic form (detritus), which has higher exergy than the elements in inorganic form. Exergy will, thereby, be a measure of complexity, but not necessarily of diversity. The exergy will generally increase as the ecological niches are better utilized by an increased biodiversity, but there may be cases, for instance, the eutrophication of an aquatic ecosystem, where the exergy increases and the biodiversity decreases (see below for further explanation).

4. Exergy can be shown by the use of statistics on modeling studies to cover a sum of buffer capacities (see below) and is thereby related to the resistance of the ecosystem.

5. Growth is increase in biomass, and the genes contain information on how to utilize the resources for growth. Evolution has continuously opened new pathways to utilize resources (including the ecological niches) better and better. Exergy considers both information and biomass measures, and therefore, the potential for growth. It is interesting in this context that there is a relationship between the exergy stored in the ecosystem and the ability of the system to capture exergy from the solar radiation (see Schneider and Kay, 1990).

6. The balance between the system components and the biodiversity are not covered by the use of exergy as a health indicator, because, for example, very eutrophic systems often have a low biodiversity and a biased distribution of the biomass, but a high exergy.

The conclusion from this comparison of Constanza's ecosystem health definition and the concept of exergy is that there is a need for supplementary ecosystem health indicators. Exergy doesn't cover all the aspects of ecosystem health presented in Constanza's definition.

Specific exergy, Exst, the exergy per unit of biomass, seems opposite Ex to be a candidate for a better coverage of points **3** and **6** in the definition of ecosystem health given above:

3. Many model studies in this volume and ecological studies (Weiderholm, 1980) show clearly that increased biodiversity means that there is a higher probability of a better utilization of the available resources, i.e., Exst increases. A better utilization of all ecological niches is accompanied by a higher biodiversity. Exst also measures the structural complexity, and the ratio of biomass to total mass (biomass + inorganic matter). A development toward a more complex organism (with more genes) will also res

6. A better utilization of all ecological niches means that there will be more species and a better balance between system components, which again may ensure a better balance between various buffer capacities.

Reactions of ecosystems to perturbations have been widely discussed in relation to the stability concepts. However, this discussion has, in most cases, not considered the enormous complexity of regulation and feedback mechanisms (see also Chapter 3).

An ecosystem is a soft system that will *never* return to exactly the same point again. It will attempt to maintain its functions on the highest possible level, but never with exactly the same biological and chemical components in the same concentrations. The species composition or the food web may have changed or may not have changed, but at least it will not be the same organisms with exactly the same properties. In addition, it is unrealistic to consider that the same combination of forcing functions will occur again. We can observe that an ecosystem has the property of resilience in the sense that ecosystems have a tendency to recover after stress, but a complete recovery, understood as being a return to exactly the same situation, will never be realized. The combination of external factors — the impact of the environment on the ecosystem — will never appear again, and even if it did, the internal factors — the components of the ecosystem — have, meanwhile, changed and can therefore not react in the same way the previous internal factors did. Resistance is another widely applied stability concept. It covers the ability of the ecosystem to resist changes, when the external factors are changed. An ecosystem will always be changed, when the conditions are changed; the question is what is changed and how much?

It is observed that increased phosphorus loading gives decreased diversity (Ahl and Weiderholm, 1977; Weiderholm, 1980), but very eutrophic lakes *are* very stable. A similar relationship is obtained between the diversity of the benthic fauna and the phosphorus concentration relative to the depth of the lakes (Weiderholm, 1980).

The concept of *buffer capacity* has a definition which allows us to quantify in modeling, and it is applicable to real ecosystems, because it acknowledges that *some* changes will always take place in the ecosystem in response to changed forcing functions. The question is how large these changes are relative to changes in the conditions (the external variables or forcing functions).

The concept is multidimensional, because we may consider all combinations of state variables and forcing functions. It implies that there are many buffer capacities corresponding to each of the state variables even for one type of change.

It was found by statistical analysis of modeling results with many different models (Jørgensen and Mejer, 1977; Jørgensen, 1992a, b, 1994a, b) that there is a correlation between exergy and the sum of a number of relevant buffer capacities. Some buffer capacities may be reduced even when the exergy increases, but it is more than compensated by an increase of other buffer capacities. These results are consistent with the relation that exergy measures the energy needed to decompose the system to inorganic components (Svirezhev, 1992). These observations explain, why it has been very difficult to find a relationship between ecosystem stabilities in the broadest sense and species diversity, as already discussed. Stability of ecosystems in its broadest ecological sense should be considered a multidimensional concept and the relation between species diversity and stability is therefore not simple and can be revealed only by a multidimensional relation. If species diversity decreases, the stability — represented by buffer capacities — may decrease in some directions, but may increase in others. It may be formulated as follows: If the system can offer a better survival, i.e., higher buffer capacities in relation to the changing forcing functions by decreasing the diversity, the system will not hesitate to react accordingly.

The above-mentioned relation between exergy and buffer capacities indicates that point 4 in the definition of ecosystem health is globally covered by the use of exergy as ecological indicator, but because there is almost an infinite number of buffer capacities, the relationship between exergy and buffer capacities can only be applied semiquantitatively in practice. As the concept of ecosystem stability, resilience, and ecosystem health are multidimensional, it will often be necessary to supplement the computations of exergy by relevant and focal buffer capacities. Buffer capacities related to the management situation should be selected. If we are concerned with the influence of toxic substances, the buffer capacities based upon the changes provoked by the input of toxic substances should be selected. If we are concerned with acid rain and its influence on the forest, we should find the buffer capacities relating the pH of rainwater to the growth of trees in the forest, and so on. The result will be a limited number of buffer capacities. To keep the ecosystem healthy, we should consider these focal buffer capacities in our environmental management strategies. As long as the buffer capacities can withstand the stress by only minor changes, the ecosystem should be considered healthy (Holling, 1992).

## 4.3   A Practical Procedure to Assess Ecosystem Health

As mentioned previously, Constanza (1992) has proposed an overall system health index consisting of system vigor, system organization, and system resilience. System vigor and the global system resilience are in accordance with the presentation above covered by the use of exergy. The organization is better covered by the structural exergy because it is highly dependent on the species diversity and their organization and independent of the total biomass concentration. As stability is multidimensional it would be an improvement in the assessment of ecosystem health to include focal buffer capacities related to actual or possible stress situations, as described above.

As pointed out by Constanza (1992) these concepts will require a heavy dose of systems modeling. It would be possible to assess the concentrations of the most important species or classes of species, and then calculate the exergy index and the specific exergy index, but it would require a dynamic model based upon mass balances to find the buffer capacities, because they relate changes in forcing functions with changes in state variables, unless it can be presumed that the relationships between forcing functions and state variables are linear. This does not imply that a new model has been developed for every new case study. Models have a certain but not complete generality. The experience from one modeling study to the next is essential. Furthermore, if the seasonal changes in exergy and structural exergy should be assessed, it will either require many measurements throughout the year or the development of a model that is able to simulate the seasonal changes. Exergy and structural exergy vary significantly during the year to reflect the ability of the ecosystem to cope with the changes in temperature, precipitation, and other climatic factors. A model will have the advantage that it can answer such questions as: How will the ecosystem health change if the forcing functions are changed so and so?

These considerations lead to the following tentative procedure for a practical assessment of ecosystem health:

1. Set up the relevant management questions related to the health of a considered ecosystem.

2. Assess the most important mass flows and mass balances related to these questions.

3. Make a conceptual diagram of the ecosystem, containing the components of importance for the mass flows defined under 2.

4. Develop a dynamic model (if the data are not sufficient, a steady-state model should be applied) using the usual procedure (see Chapter 3 and Jørgensen, 1994a).

5. Calculate exergy, specific exergy, and relevant buffer capacities by the use of the model. If the model is dynamic, it will also be possible to find the seasonal changes in exergy, structural exergy, and buffer capacities.

6. Assess the ecosystem health: high exergy, specific exergy, and buffer capacities imply a good ecosystem health. If the exergy and specific exergy is high but one of the focal buffer capacities is low, the remedy is to improve the structure of the ecosystem to assure a higher focal buffer capacity. If the exergy is high but the specific exergy and some focal buffer capacities are low, we would probably be dealing with a stressed, e.g., a eutrophic system, where the remedy should be reduction of the nutrient loadings. Based upon the values of the three indicators, different measures should be taken to improve the ecosystem health.

The procedure should be considered a first tentative approach to assessment of ecosystem health. Most probably more indicators should be included when more experience in assessment of ecosystem health has been attained. It is not surprising if 10 or even more indicators are needed to do a proper assessment of ecosystem health. The experience gained by using this procedure shows, however, that the proposed steps are applicable in a practical environmental management context.

## 4.4    Assessment of Ecosystem Health Exemplified by Lakes

Illustrations of the application of exergy, structural or specific exergy, and buffer capacity as indicators of ecosystem health are given below for lake and agricultural systems. The examples are reproduced from Jørgensen, 1994a and Jørgensen, 1997. The same approach has been applied on models of wetlands, fishponds, streams, coastal ecosystems, and agricultural systems with the same general results.

A eutrophication model with seven state variables has been applied: nutrients, phytoplankton, zooplankton, planktivorous fish, carnivorous fish, detritus, and sediment. The usual equations (see Jørgensen, 1976, 1994a, 1997; Jørgensen et al., 1978) are applied, but have been expanded with the following characteristics according to ecological observations:

1. Threshold concentrations are used for grazing and predation. Below these, no grazing and predation take place.

2. Predation by the most carnivorous fish on the planktivorous fish is reduced above a certain concentration of phytoplankton because carnivorous fish are hunting by sight.

3. The growth rate of phytoplankton and zooplankton is reduced step-wise from low nutrient concentrations to high nutrient concentrations. This is in accordance with the observations that bigger species prevail at a higher level of eutrophication, as discussed above.

4. Adaptation to changed temperature is used by changing the optimum temperature accordingly.

5. The flow rate relative to the volume is 0.1, which assures a fast reaction to the nutrient concentration in the in-flowing water.

6. The model does not distinguish between phosphorus and nitrogen, but assumes that they are present in the ratio used by phytoplankton, i.e., 1:7. The photosynthesis follows the uptake of nutrient by a factor of 12, corresponding to 11 times as much uptake of carbon, hydrogen, oxygen, and other elements than of nitrogen and phosphorus.

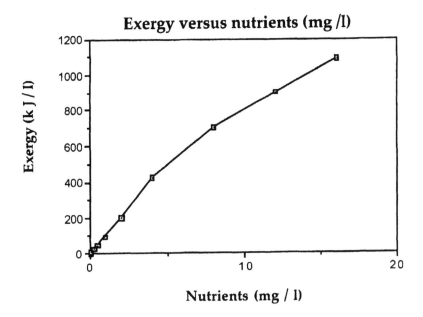

**FIGURE 4.2**
The results obtained from a eutrophication model with seven state variables are shown. The exergy is plotted vs. the total inputs of nutrients (nitrogen and phosphorus).

The model computes the exergy and the structural exergy. The buffer capacities for changes in phytoplankton, zooplankton, and the two classes of fish, when the inputs of nutrients and the temperature are changed, are found by the use of a sensitivity analysis.

The results are shown in Figures 4.2 through 4.4. Exergy, specific or structural exergy, and buffer capacity of phytoplankton to changed nutrient loading are plotted vs. the nutrient concentration in the in-flowing water. As seen, the exergy increases with the increased nutrient input due to the resulting higher total biomass concentration. Structural exergy has a maximum at the nutrient concentration of about 2 mg/l. By higher nutrient inputs, the structural exergy declines due to an unequal distribution of the biomass. Particularly the phytoplankton and the planktivorous fish increase on behalf of zooplankton and carnivorous fish. A structural change is observed, which is consistent with general observations in lakes (see also, Jørgensen, 1995, 1997). The buffer capacity, phytoplankton to changed nutrient loading, has a minimum total nutrient input of about 2 mg/l. It increases by increased nutrient loading, mainly due to the slower growth rate.

Other changes in buffer capacities can be summarized as follows:

1. The buffer capacity for the influence of nutrients on the carnivorous fish increases with increasing nutrient loading. After a certain nutrient input, the concentrations of carnivorous fish remain low at almost the same level, independent of the nutrient concentration.

2. The buffer capacities for the influence of nutrients on zooplankton and planktivorous fish have a maximum at 1 mg/l and are decreasing above this concentration of nutrients.

3. The buffer capacities for the influence of temperature generally decrease with an increased nutrient input above a nutrient concentration of about 2 to 4 mg/l except for carnivorous fish, where the buffer capacity consequently increases slightly with increasing nutrient input.

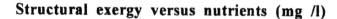

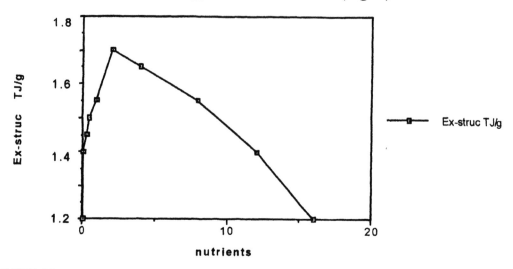

**FIGURE 4.3**
The results obtained from the use of a eutrophication model with seven state variables are shown. The structural exergy is plotted vs. the total inputs of nutrients (nitrogen and phosphorus).

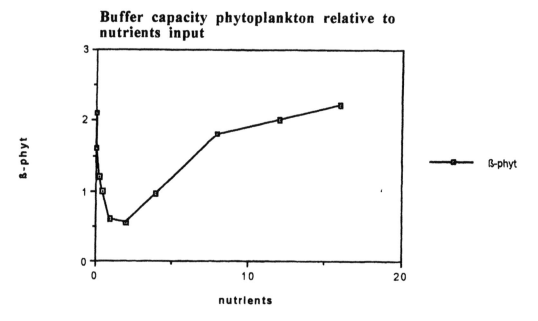

**FIGURE 4.4**
Buffer capacity of phytoplankton to changed nutrient loadings (mg/l) are plotted versus the nutrient concentration in the in-flowing water.

The results are consistent with the general observations and previous model studies (see Jørgensen, 1976, 1994a, 1997; Jørgensen et al., 1978). The buffer capacities are generally either increasing or constant up to a nutrient level of about 1 to 4 mg/l, except for the influence of nutrients on phytoplankton. At about the same concentration, structural exergy has its maximum.

The results may be interpreted as follows: up to a total nutrient concentration of about 2 mg/l, the examined buffer capacities are rather constant or even increasing, except for the influence of nutrients on phytoplankton. Structural exergy, which measures the ability of the system to utilize the resources, increases along the same line.

In this range of nutrient loading, the top-down and the bottom-up controls are working in parallel (see Sommer, 1989). A nutrient loading up to this level seems fully acceptable, but if the loading increases above this level, crucial buffer capacities and the structural exergy decrease.

Some buffer capacities measuring the influence of nutrient input on phytoplankton and carnivorous fish increase with increasing nutrient concentrations. It reflects the ability of the system to meet changes in forcing functions, in the case of the nutrient loading, by such changes in the structure, that the direct influence of these changes is reduced (in this case the influence of nutrients on phytoplankton). In this range of nutrient loading the top-down control has collapsed (see Sommer, 1989). Notice also that the exergy increases with increasing nutrient concentrations. A nutrient concentration above approximately 2 mg/l should accordingly be omitted, and measures should be taken to reduce the nutrient loadings to this level or below.

The results of an ecosystem health analysis, as illustrated in this case study, can only be interpreted semiquantitatively. This is because the accuracy of the underlying model does not allow very precise quantifications. The case study has, on the other hand, shown that the analysis is very useful, because it enables us to approximate the acceptable level of nutrient loadings and to see the consequences in relation to buffer capacities, if we increase the loading above this level, i.e., to predict the expected outcomes of structural changes in the system.

In all, 15 lake case studies taken from ICLARM's ECOPATH survey on various ecosystems (Christensen and Pauly, 1993) were compared with relation to exergy and structural exergy. The lakes are taken from the steady-state average situations based upon observations. The exergy increases with increasing eutrophication for the 15 lakes, as shown in Figure 4.5, where the exergy is plotted vs. the total biomass of phytoplankton and macrophytes, including export. Approximately the same picture and trends as obtained by the eutrophication model are seen (compare Figures 4.2 and 4.5). Figure 4.6 shows from these 15 case studies structural exergy = exergy divided by the total biomass vs. the eutrophication, measured by the biomass of phytoplankton and macrophytes, including export.

Structural exergy has a maximum at a relatively low level of eutrophication, which is to be compared with a medium level at the model exercise (see Figure 4.3), as most of the 15 examined lakes are eutrophic to hypereutrophic. Figures 4.5 and 4.6 indicate that the results found by modeling studies, shown in Figures 4.2 and 4.3, are also qualitatively valid, when direct lake observations are used.

## 4.5   Sustainability of Agroecosystems

There is a growing concern that modern agriculture destroys it own resource base. The intensive use of external agrochemical inputs in the displacement of many natural biological and ecological processes and functions in ecosystems is probably at the root of this concern.

Many modern agroecological systems are characterized by extensive dependence and impact on environment, which means that they have large inputs and outputs. Agricultural

**Exergy / total biomass incl. export**

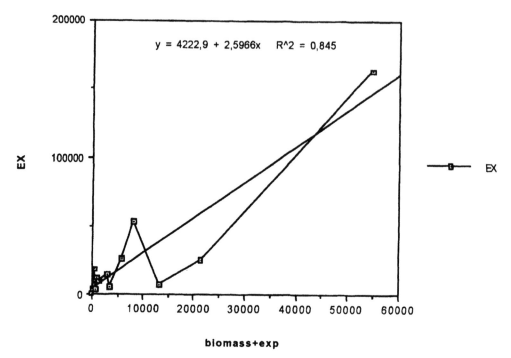

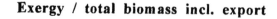

**FIGURE 4.5**

Exergy in J/l is plotted for 15 lake studies. Taken from Christensen and Pauly, 1993. The exergy increases with the biomass + export (μg/l), as expected from Figure 4.2.

activities are thus gradually removing the inherent ability of the underlying ecosystem to sustain and regenerate itself (see Dalsgaard, 1996).

It would be obvious to attempt to apply the ecological indicators presented in this chapter on agroecological systems. Dalsgaard (1996, 1997) has used a steady-state model, ECO-PATH II (see Dalsgaard and Oficial, 1995), to carry out such an analysis on the basis of four different tropic farms, Farms A–D. The results are summarized in Table 4.2. Farm A is a monoculture rice agroecosystem with four state variables: phytoplankton, rice, maize, and grass (weeds). Farm B includes aquaculture and has three additional state variables: azolla, fruit trees, and fish. Vegetables have replaced maize. Farm C is a diversified and integrated rice-based agroecosystem. It has, in addition to the four state variables of Farm A, seven state variables: fruit trees, multipurpose trees, bamboo, fish, poultry, ruminants, and pigs. Vegetables have also here replaced maize. Farm D has the same state variables as Farm C, but uses composting to obtain a higher extent of mass cycling.

It is the general perception that the extent of integration and the maturity of the four farms increases from A to D. Structural exergy, the Shannon index, Finn's cycling index, and H/E all follow roughly the same trends, while P/B, not surprisingly, decreases from A to D. The biomass is dominated by the presence of trees, which is also reflected in the exergy. Trees may be important for the microclimate (lower temperatures during the day and higher temperatures at night). Trees will, furthermore, due to their high biomass be able to utilize (capture) the incoming solar radiation better. It may, therefore, be important to ensure a high exergy in addition to the indices expressing a high utilization of the available resources (structural exergy), a high diversity (Shannon's index), a high cycling (Finn's cycling index), and a high yield relative to the throughput (H/E).

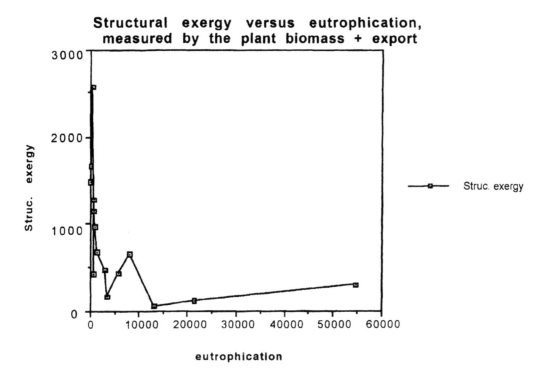

**FIGURE 4.6**

Structural exergy (kJ/g) is plotted vs. the eutrophication for 15 lake studies.

**TABLE 4.2**

Comparison of Ecological Indicators and Attributes for Four Tropic Farms (the differences are explained in the text)

| Indicator | Farm A | Farm B | Farm C | Farm D |
|---|---|---|---|---|
| Biomass (kg/ha/y) | 8080 | 44300 | 16500 | 38700 |
| Harvest (H) | | | | |
| kg N/ha/y | 87 | 163 | 98 | 160 |
| Exergy* | 468,000 | 2,580,000 | 1,070,000 | 2,550,000 |
| Struct. exergy* | 58 | 58 | 65 | 66 |
| Shannon index | 0.7 | 1.0 | 1.6 | 1.6 |
| Finn's cycling index** | 0.29 | 0.43 | 0.52 | 0.44 |
| P/B | 2.8 | 1.5 | 1.3 | 0.7 |
| Efficiency# | 0.33 | 0.27 | 1.36 | 0.76 |
| N throughput (E) | | | | |
| kg N/ha/y | 624 | 687 | 319 | 312 |
| H/E | 0.14 | 0.24 | 0.31 | 0.51 |

\* Exergy is indicated as kg detritus equivalent/ha/y, and structural exergy as exergy/ biomass.

\*\* Finn's cycling index expresses the recycled fraction of total throughput, based on the nitrogen flows.

\# Efficiency is defined as the output (harvest)/inputs (fertilizers and feeds).

Agroecosystems are complex systems. It is, therefore, recommended to apply several indices simultaneously to get a proper image of the extent of integration, the maturity, and perhaps also the sustainability of agricultural systems. All the indices included in Table 4.2 tell their own story about the system.

The biomass and the exergy are direct measures of the maturity according to Odum (1969), but a high exergy should be preferred as an index because it measures directly the (relative) distance from thermodynamic equilibrium and is related to the sum of buffer capacities. The agroecosystem will generally be closer to the natural ecosystem, the lower the P/B ratio and the higher the biomass are.

Structural exergy measures how the available resources are utilized to construct a system as far from thermodynamic equilibrium as possible, as it is calculated by exergy relative to the biomass.

The amount of harvest gives the yield, which is the ultimate goal for farms to maintain on as high a level as possible on a long-term basis. The efficiency gives the yield relatively to the inputs. Efficiency should be preferred to a high yield, measured as kg/ha/y. The latter gives the intensity of the agriculture, while a high efficiency is more related to the sustainability — a continuous long-term harvest independent of inputs. H/E may be considered a measure of sustainability, because it measures the harvest relative to the throughput. This ratio also follows the trends in structural exergy, which as presented above, may be considered an expression of the utilization of resources. Utilization of the available resources seems also to be a good approach to a sustainability measure. The cycling index and the Shannon index give additional direct measures of the organization and the diversity, but further studies based on a wider range of agricultural systems are needed.

---

## 4.6 Overview of Ecological Indicators

Different goal functions, ecological indicators, and orientors have been proposed in the ecological literature, but because they all aim toward an expression of the information embodied in the ecological network, it is not surprising that they are strongly correlated. Exergy measures the information of the system, based on the genes of the organisms that are forming the ecological network. Structural or specific exergy is exergy relative to the total biomass, and should therefore be related to the information in the network, independent of the size of the network = biomass represented by the network. Emergy expresses the amount of solar energy that it has required to construct the network. Ascendancy, A, looks into the complexity of the network itself, including the amount of mass or energy that is flowing through the network. Details about these concepts can be found in Jørgensen, 1997. The ratio of indirect to direct effect, denoted I/D, measures the importance of cycling relative to the direct cause, and this ratio is therefore also a measure of the information of the network. The different concepts give slightly different information and are deducted from different viewpoints, but they all contribute to an indication of the integrity of the focal ecosystem.

Particularly, the specific exergy and the indirect effect/direct effect ratio, when ecosystems (or models of ecosystems) with *different complexity* are compared, and exergy and ascendancy, for different flow situations in the *same* ecosystems (models), can be shown to be well correlated. Exergy and ascendancy measure both the complexity, the size, and the level of information of the system, although the latter is based on the flows and the former on the storages. The strong correlation between the two concepts, confirmed a few times throughout the system ecological literature, is therefore explainable (see Jørgensen 1994b, 1997).

**TABLE 4.3**

Overview of Applicable Ecological Indicators for an Extensive and Intensive Description of Ecosystems under Development and of Mature Ecosystems

| Ecosystem Stage | Extensive Description | Intensive Description |
|---|---|---|
| **Under Development** | Exergy captured | Structural exergy |
| | Exergy stored | Overhead |
| | Ascendancy | Production/biomass |
| | Biomass | Cycling index |
| | Indirect effect | Indirect effect/direct effect |
| | Emergy | A–T/T |
| | | Emergy/exergy |
| **Mature** | Exergy stored | Structural exergy |
| | Ascendancy | Overhead |
| | Minimum excess entropy | Entropy production/biomass |
| | Indirect effect | Indirect effect/direct effect |
| | Emergy | A–T/T |
| | | Emergy/exergy |

Specific exergy and I/D for different ecosystems (models) measure the complexity of the structure of the network, independent of the size (the total biomass or the total through-flow of energy). It is therefore not surprising that they are strongly correlated.

Since emergy measures the cost in solar equivalents and exergy the result of the solar radiation stored in the ecosystem as biomass, structure, and information, it is not surprising that the ratio of emergy/exergy may be a strong indicator of ecosystem development, as was found by Bastianoni and Marchettini (1997). The mature ecosystem will offer a high efficiency in its use of solar radiation to maintain and store the ecological structure. They found that the ratio was 37 to 84 for a waste pond, while it was 0.22 to 0.24 for a natural lagoon. The latter is about 200 times more effective in transferring solar radiation into stored biomass and information. The natural ecosystem will move 200 times farther away from thermodynamic equilibrium by utilization of the same amount of solar energy, compared with the man-made waste pond system.

Table 4.3 attempts to summarize the experience gained by application of various proposed ecological indicators. For ecosystems at an early stage (under development), and at a mature stage (developed), applicable indicators are proposed for a description of the integrity of ecosystems. Both an extensive description considering the amount of the available resources (includes size terms) and an intensive description accounting for how well the ecosystem is able to utilize the available resources are taken into account in Table 4.3.

# References

Ahl, T. and Weiderholm, T., 1977, Svenska vattenkvalitetskriterier: Eurofierande ämnen. Svenska Naturritens Kaps Akademi, Sweden.

Bastianoni, S. and Marchettini, C., 1997, Comparison of emergy and exergy for three lagoons, *Ecol. Modelling*, 62, 33.

Christensen, V. and Pauly, D., 1993, *Trophic Models of Aquatic Ecosystems*. ICLARM. International Council for the Exploration of the Sea, Danida.

Constanza, R., 1992, Toward an operational definition of ecosystem health, in Constanza, R., Norton, B.G., and Haskell, B.D., Eds., *Ecosystem Health. New Goals for Environmental Management*, Island Press, Washington, D.C.

Dalsgaard, J.P.T., 1996, An ecological modelling approach towards the determination of sustainability in farming system, thesis, KVL, Copenhagen.

Dalsgaard, J.P.T., 1997, Agroecological sustainability and ecosystem maturity, *Ecol. Modelling*, in press.

Dalsgaard, J.P.T. and Oficial, R.T., 1995, Insights into the ecological performance of agroecosystems with ECOPATH II. NAGA, *The ICLARM Quart.*, 18, 26.

Dalsgaard, J.P.T. and Oficial, R.T., 1998, Modeling and analyzing the agroecological performance of farms with ECOPATH, ICLARM Technical Report 53, 54.

Holling, C.S., 1992, The role of forest insects in structuring the boreal forest, in Shugart, H.H., Ed. *A System Analysis of the Global Boreal Forest*, Cambridge University Press, Cambridge.

Jørgensen, S.E., 1976, A eutrophication model for a lake, *Ecol. Modelling*, 2, 147, 165.

Jørgensen, S.E., 1992a, Development of models able to account for changes in species composition, *Ecol. Modelling*, 62, 195.

Jørgensen, S.E., 1992b, Parameters, ecological constraints and exergy, *Ecol. Modelling*, 62, 163.

Jørgensen, S.E., 1994a, *Fundamentals of Ecological Modelling*, 2nd ed., *Developments in Environmental Modelling*, 19, Elsevier, Amsterdam.

Jørgensen, S.E., 1994b, Models as instruments for combination of ecological theory and environmental practice, *Ecol. Modelling*, 75/76, 5.

Jørgensen, S.E., Mejer, H.F., and Friis, M., 1978, Examination of a lake model, *Ecol. Modelling*, 4, 253.

Jørgensen, S.E., 1997, *Integration of Ecosystem Theories: A Pattern*, Kluwer, Academic Publishers, Dordrecht.

Kay, J.J., 1991, A non-equilibrium thermodynamic framework for discussing ecosystem integrity, *Environ. Manage.*, 15, 483.

Kay, J.J. and Schneider, E.D., 1990, On the applicability of non-equilibrium thermodynamics to living systems [internal paper], University of Waterloo, Ontario, Canada.

Kay, J.J. and Schneider, E.D., 1992, Thermodynamics and measures of ecological integrity, in *Ecological Indicators*, Elsevier, Amsterdam, 159.

Odum, E.P., 1969, The strategy of ecosystem development, *Science*, 164, 262.

Rapport, D.J., 1995, Preventive ecosystem health care: the time is now, *Ecosystem Health*, 1, 127.

Schneider, E.D. and Kay, J.J., 1990, *Life as a Phenomenological Manifestation of the Second Law of Thermodynamics*, Environment and Resource Studies, University of Waterloo, Ontario, Canada.

Svirezhev, Y., 1992, Exergy as a measure of the energy needed to decompose an ecosystem, presented at ISEM's (International Society of Ecological Modelling) International Conference on the State-of-the-Art of Ecological Modelling, 28, September–2, October 1992, Kiel.

Weiderholm, T., 1980, Use of benthos in lake monitoring, *J. Water Pollut. Control Fed.*, 52, 537.

Wulff, F. and Ulanowicz, R.E., 1989, A comparative anatomy of the Baltic Sea and Chesapeake Bay ecoystems, in Wulff, F., Field, F.G., and Mann, K.H., (Eds.), *Flow Analysis of Marine Ecosystems*, Springer Verlag, Berlin.

# 5

## Environmental Management System

**Rikke Dyndgaard and John Kryger**

## CONTENTS

## 5.1 Why Introduce an Environmental Management System?

Environmental management is a systematic approach to environmental care in all aspects of business. Implementation of this approach is normally a voluntary process. Companies are, however, not only assessing benefits of adopting environmental management in financial terms (savings, production efficiency, market potential) but also the risks of not adequately addressing environmental aspects (accidents, inability to obtain bank credit and other investment money, loss of markets).

## 5.2   Introduction

The health and well-being of people and the environment depends on what people do today. Unless people make drastic changes in the way they live, work, and play, the Earth will continue to suffer; the environment on which they depend for survival will continue to deteriorate.

Industry can be a source of some of the most serious local and international environmental problems, particularly those related to pollution of the air, water, and soil. These problems can have rapid or gradual impacts on human health, affecting entire neighborhoods, cities, or regions of a country. Pollution can travel through the air, rivers, and groundwater from one country to another, having severe effects on the quality of life of people living downwind or downriver from the source of the pollution.

The business community has realized that the current patterns of production and consumption áre unsustainable. At the same time, enterprises have realized that, in order to stay in business, they will increasingly have to integrate environmental considerations into their business strategy and long-term planning. This is essential if they are to use business opportunities, compete with other businesses that take environmental considerations into account, and provide for interested parties/stakeholders with increasing environmental expectations.

An environmental strategy and policy is, of course, a starting point for businesses to integrate environmental aspects into their operations. Tools to ensure systematic attention and the achievement of policy and objectives include, among others, EMS and environmental auditing. These help to control and improve the environmental performance in line with the company environmental policy. Additional tools could also be environmental life cycle assessment methodologies, environmental labeling programs, and performance evaluation methods. These tools will not be described in this paper.

These instruments have been promoted in several countries by the governmental bodies as suitable instruments for companies to adopt and use at their own discretion, without legislative pressure. At the same time business and other organizations at the international and national levels have also promoted these instruments as useful tools. In applying these instruments, many companies and their interested parties have encountered the need for more clarity on the details of EMS and environmental auditing concepts, and at the same time the need emerged for a level playing field in relation to these aspects. Activities on standardization and certification have therefore started at both the national and international levels. The International Organisation for Standardisation (ISO) has developed standards for EMS, environmental auditing and several other environmental management tools.

## 5.3   International Developments since the 1970s

In Western Europe and the United States, and increasingly evident worldwide, the approach to controlling or influencing the impact of industrial activity on health and the environment has undergone a significant transition. Initially (in the 1970s and early 1980s in Europe) efforts concentrated on the development of legislative and regulatory structures together with enforcement through an environmental permit structure. The response of industry was largely reactive. Industry invested in "end-of-pipe" technological solutions to ensure compliance with the ever-increasing regulations and the environmental conditions attached to operating permits.

The combination of business and environmental aspects in the international arena started after the 1972 United Nations Conference on Human Environment when an independent commission was created: the World Commission on Environment and Development (Brundtland Commission). This commission took up the task of reassessing the environment in the context of development and published its report, *Our Common Future,* in 1987. It is now considered to be a landmark report. It introduced the term *sustainable development*, and urged industry to develop effective EMS. It was adopted by more than 50 world leaders, who called for a major conference for further discussion and decisions.

The UN therefore decided to organize the UN Conference on Environment and Development (UNCED), also referred to as the Earth Summit, held in Rio de Janeiro in June 1992. Government leaders, business leaders, and private groups met to consider how the world can move toward sustainable development.

The outcome of the Earth Summit conference was Agenda 21, a "global consensus and political commitment at the highest level" on how governments, enterprises, nongovernmental organizations, and all sectors of human life and action can cooperate to solve the crucial environmental problems of our time which threaten human life and society.

The Secretary General of UNCED wanted to ensure that business would participate in the process of discussion and decision making. He therefore asked a leading Swiss industrialist to be his principal adviser on business issues. This industrialist performed this role by establishing the Business Council on Sustainable Development (BCSD). This BCSD published its important report, *Changing Course*, but also decided to approach the ISO to discuss the development of environmental standards.

In parallel with these developments, the International Chamber of Commerce (ICC) developed the Business Charter for Sustainable Development in 1990, which was launched the following year at the Second World Industry Conference on Environmental Management (WICEM). The ICC Business Charter contains 16 principles of sound environmental management.

In another initiative, the chemical industry, concerned about its deteriorating public image, launched its Responsible Care Programme, beginning in Canada in 1984 and now a condition of membership of the Chemical Industries Association. Its approach is based firmly on the principles of total quality management, including assessment of actual and potential health, safety and environmental impacts of activities and products, and the provision of information to interested parties.

Since the mid-'80s in the West, and more recently in the emerging and dynamic economies of the East and the West, industry is taking a more proactive stance and is recognizing that sound environmental management on a voluntary basis can enhance corporate image, increase profits and competitiveness, reduce costs, and obviate the need for further legislative measures by the authorities. Evidence of this is seen in the move toward "green products" with the increasing use of "life cycle analysis" — looking at the environmental impact of a product "from cradle to grave." It has also produced a number of environmental management tools, such as environmental auditing and EMS. These tools largely started as voluntary in-company initiatives but are now affecting the policies and regulatory approach of the European Union, governments, and the risk management policies of national and international banks and insurance companies.

In most countries, the implementation of structured environmental management in companies remains voluntary. Companies worldwide, however, are carefully assessing not only the potential financial benefits (identification of savings, improved production efficiency, new market potential, etc.) which may arise from such activities, but also the risks of not addressing organizational as well as technical solutions to environmental problems (accidents, inability to obtain bank credit and private investment, loss of markets and customers).

One of the most important activities of the last few years is perhaps the development of standards in the environmental field, especially those being undertaken by the ISO. These are essential if EMS (and related activities) is to be applied within the context of the "level playing field" as required by international trade agreements both within the EU and worldwide (GATT/WTO). Also standard developments at the national and European levels are affecting industry worldwide, the main developments being the recognition of the British Standard for EMS (BS 7750) in many countries and the implementation of the Eco Management and Audit Scheme (EMAS) in the European Union.

### 5.3.1   Danish Development

Danish enterprises have shown an increasing interest in implementing EMS during the last nine years. At the beginning of the 1990s the first Danish enterprises were certified after the British Standard, BS 7750. In 1993 came the EMAS, which led to the first EMAS registration in Denmark in 1995. The EMAS is often used by enterprises in one EU country that exports to another EU country. In 1996 it became possible for Danish enterprises to be certified according to ISO 14001. After the launching of ISO 14001, the BS 7750 became irrelevant, and after 1997 it has not been possible to be certified according to BS 7750. Besides the well-known standards (BS 7750, EMAS, and ISO 14001), many enterprises have chosen to develop and implement EMS tailored to their specific needs. This means that some enterprises have EMS that has not been certified or registered. At the moment (beginning of 1999), about 50 enterprises have an EMAS registration, and 200 have an ISO 14001 (or BS 7750) certificate.

## 5.4   Environmental Management

Environmental management can help companies approach environmental issues systematically and integrate environmental care as a normal part of their operations and business strategy. Some major triggers for companies in this context are:

- Legislation and enforcement:
  - an increasing volume of policies, laws, and regulations, and their enforcement
- Stakeholder pressure:
  - the increasing pressure of third parties, such as financial institutions and insurance companies
  - pressures from stakeholders and employees
  - attention from environmental interest groups, consumers and their organizations, and the general public (in the locality)
- Awareness, image, and reputation:
  - growing awareness in the business community concerning the environment (responsible care); corporate image (public, authorities)
  - some of the impacts on business of accidents and failures in environmental management controls (negative publicity, damage to corporate image)

- Competitiveness
  - growing awareness that the environmental aspects of products and processes may play a role in international competitiveness
  - the fear of international trade barriers formed by different standards for environmental performance
- Finance
  - some of the impacts on business of accidents and failures in environmental management controls (liability issues, cost of remediation, business interruption)
  - the introduction of economic (financial) instruments, such as taxes or levies on emissions (e.g., waste), to stimulate a decrease in pollution levels
  - incentives from government (licensing), banking (more attractive credit facilities), and insurance companies (more attractive premiums)
  - cost savings through cleaner production and eco-efficiency

### 5.4.1 Legislation and Enforcement

Governments at all levels are strengthening the control of industrial activities and increasing the penalties for violation of environmental regulations and permits. Judiciaries in developing countries in particular have, in recent years, become very "green" indeed. New and more serious civil and criminal penalties are imposed on the violation of environmental permits and laws, particularly if the violation leads to a health risk or to the long-term damage of natural resources, such as groundwater or soil quality. Enterprises are required to carry out the monitoring necessary to prove that they are in compliance with their permits and with applicable legislation.

Civil or criminal liability for violations of laws and administrative regulations is becoming more severe, and the basis of liability is being broadened in many countries to cover any environmental damage without evidence of fault ("strict" liability). The related business risks include the loss of freedom to operate or produce and the imposition of civil or criminal penalties directly on top management.

The authorities may increase the level of control and be less agreeable to negotiation and compromise with an enterprise that is known to have problems complying with environmental laws and permits, than with an enterprise with a good compliance record.

Apart from criminal and civil liabilities for damages from accidental or other pollution, enterprises in a growing number of countries are also facing emergency abatement notices, where authorities order them to stop production until the emergency has passed. In some cases, industries have been forced to relocate or to make substantial capital investments in new pollution control equipment. In other cases, enterprises have been forced to pay for cleaning up the problem.

### 5.4.2 Stakeholder Pressure, Awareness, Image, Reputation

Enterprises are expected to be good citizens, and there is growing peer pressure to maintain certain minimum standards and an accepted level of vigilance to guard against environmental problems. Many people believe that an enterprise with poor environmental performance cannot possibly produce high-quality products. Consumers increasingly favor products that are produced under conditions that are viewed as less burdensome on the environment.

Other stakeholders such as financial institutions and insurance companies, increasingly evaluate the environmental performance of an enterprise in the course of the overall evaluation as a potential or an existing client, before further services are rendered or favorable conditions can be negotiated.

An important "customer" of every enterprise is the local community where it operates. The most successful companies in industry today take the view that they must "earn the right" to operate in each community where they are present. These enterprises raise the standards of environmental performance for their competitors.

Enterprises that disregard the concerns of local populations with respect to pollution sometimes find their applications for permits, and their entrance gates, blocked by protesters. In one extreme case, a crowd of angry citizens in Thailand actually attacked and burned a factory to the ground over an incident of pollution of drinking water. Usually, however, opposition results in costly delays in obtaining planning and construction permission, and difficulties with regulatory authorities.

Environmental and consumer groups are becoming more influential in many countries, as is press coverage. All of these can affect an enterprise's image throughout the country and abroad, as well as its ability to sell goods and obtain loans or capital investment. Environmental disputes can not only harm an enterprise's reputation, they can turn the attention of managers and workers away from their main responsibilities, they can lower morale, and they can bring the enterprise into a state of uncertainty and confusion.

"Clean" companies are frequently perceived as good neighbors and responsible firms that inspire the confidence of authorities and customers. They are more likely to be consulted by policy makers who are preparing new legislation. They may have a stronger negotiating position with the permit authorities in case of accidents or the agreement of pollution control or environmental improvement plans.

As the environmental performance of the enterprise improves, it can publicize its progress to increase its value and possibly its market share. For example, some companies have public environmental policies, action plans, or "Green Accounts." They document their progress toward reducing pollution and waste and toward meeting their environmental goals and targets. Other enterprises publicize their use of recycled materials or of cleaner technology. Others develop products that use recycled or biodegradable materials, or meet customer demands for less hazardous wastes, and market their products as "green."

### 5.4.3   Competitiveness

An enterprise can lose its competitive position in domestic as well as international markets by failing to pay attention to environmental issues. The most obvious way is through the higher costs which can come from wasted materials and energy. An enterprise that fails to take the environment into account can produce lower-quality products which may be rejected by customers. Exposure to wastes and pollution may cause injury or illness to workers or to the local community.

Today in the United States, the European Union (EU), and other affluent areas of the world, enterprises must assume environmental responsibility, and in many cases accept liability for the environmental impacts of their actions, as a matter of basic business practice. Until recently, this trend has not applied to enterprises in the emerging and industrializing economies. However, rapidly evolving global business practices and international agreements are set to change this. Enterprises with serious environmental liabilities may be driven out of business.

"Green consumerism" is now a significant market force. Producers or firms that sell to consumers frequently demand that their suppliers meet newer, tougher product environmental criteria. Such demands can range from assurances that the products supplied meet all legal requirements in the country to which they are exported to demands that suppliers meet certain minimum environmental criteria in their own business and manufacturing practices. High standards for the quality of a product may mean that a supplier must impose equally high standards on the production process. Some companies may impose process standards directly on their suppliers, and monitor compliance with their standards.

Laws may be passed by parliaments many thousands of kilometers away which have no direct jurisdiction over the enterprise, but which — by banning a product or product component — can put an enterprise out of business. As many formerly acceptable substances are being regulated out of products because of their health or environmental implications, suppliers of products containing such substances are forced to find alternatives or face loss of markets. Domestic or foreign competitors may be gaining a competitive advantage with consumers by selling the environmental "friendliness" of their products or production processes. In this case an enterprise may be obliged to prove a certain level of environmental compliance to the public if it wants to remain competitive.

The owner or manager of an enterprise should be concerned with environmental management for one fundamental reason: Poor environmental results will decrease the value of their activity and undermine their competitive advantage.

At first glance, environmental concerns may seem irrelevant to companies operating in today's emerging market economies, whose main priority is survival and rapid liquidity. This seems even more true for small and medium-sized enterprises that sell products or services to a mainly national market.

There is a growing web of national and international standards and requirements that enterprises will have to comply with if they want access to new international markets, or even if they want to maintain good relations with the enterprises and countries to which they already are exporting their products.

In the very near future, enterprises may not be able to export products that are dangerous to the environment (low biodegradability, containing dangerous substances, etc.) or that are made under "unacceptable" environmental conditions. Efforts have been made to reduce emissions across many industries, for example, and the list of regulated products and wastes is steadily growing.

Enterprises in emerging economies soon will have to play by the same rules as their counterparts in the U.S. and the EU. To be able to sell their products in the U.S. and EU, they will have to demonstrate that they follow accepted international environmental practices.

Some governments are adopting legislation stimulating (not requiring) enterprises to audit the environmental impacts of their activities and to reduce waste or improve their technologies. Others, such as the EU, have adopted broad requirements for environmental management and audit schemes. The ISO has developed a standard (among others) for EMS, ISO 14001. (The new World Trade Organisation [WTO] will pay increasing attention to the environmental aspects of trade in the coming years, as environmental standards and business practices become more international.)

Environmental laws are becoming more harmonized in all parts of the world. In the near future, the laws in "high-tolerance areas" will be brought up to a world standard. The free trade areas such as the EU, European Economic Area, and NAFTA in the Americas, have common structures for environmental regulation, compliance, and policing. The emerging economies in South America, Africa, Asia, eastern Europe, and the new independent states will be asked to meet the same standards if they wish to export to these regions.

Enterprises around the world also will be subject to an increasing number of international environmental conventions in the very near future. For example:

- Traders in wastes must comply with the Basel Convention on the control of Transboundary Movements of Hazardous Wastes and their disposal.
- Energy enterprises in Europe will have to comply with the limits on sulfur emissions, nitrogen oxides, and volatile organic compounds under the ECE Convention on long-range transboundary air pollution.
- Enterprises selling chemicals to the EU will have to comply with regulations on the import, packaging, and labeling of hazardous chemicals.

At the production stage, the technical improvements in environmental performance can sometimes be developed as new products for new markets. Existing products may benefit from environmental sensitivities in other countries and find new markets.

Products that are more responsive to the environmental sensitivities of customers may have an important competitive advantage over those of competitors who do not attach importance to these considerations, particularly if the price difference is reasonable.

### 5.4.4   Finance

Enterprises that find ways of reducing or even eliminating pollution, waste, and energy consumption can bring about important cost savings, and thus be more competitive. Common cleaner production opportunities are found in areas such as energy efficiency, emission reductions, recycling, or recovering value from waste, minimizing raw material usage, etc.

This behavior of enterprises is often encouraged by economic incentives, such as taxes or levies on emissions and waste.

Banks will increasingly review the environmental liabilities of a company before securing the capital asset for financing; similar trends are occurring with insurance companies. An enterprise with lower environmental risks can obtain project financing more easily, at more favorable conditions, and insurance at lower costs.

Tougher legislation can also lead to higher production costs, not only because of new capital investments required of the enterprise, but also because of new requirements imposed on the enterprise's suppliers, including the power company. Enterprises which are in a position to foresee the impacts of regulatory developments on their activities are in a position to reduce the costs of compliance.

An enterprise without major environmental problems has staff who are free to concentrate on their main tasks without fear of major disruption from accidents or public protests about the environmental impacts from the enterprise's activities. The avoided burden of disrupted operations and low staff morale should not be underestimated.

## 5.5   Environmental Management Systems

Companies therefore try to achieve and demonstrate sound environmental control and active management of their environmental performance. In many companies there is already some tradition of performing environmental "audits." These investigations, although they are useful in environmental management, cannot provide the assurance that the company's environmental performance meets, and will continue to meet, the requirements of the company environmental policy and relevant legislation. It is now accepted that a better approach is to develop, implement, and maintain a well-structured EMS, integrated with overall management activity and addressing all aspects of desired environmental performance.

At present there is worldwide attention for EMS in companies. EMS offer a structured and systematic method to incorporate environmental care in all aspects of business. The aim is not only to comply with environmental regulations and minimize the (financial) risks of liabilities and costs, but to improve the environmental performance continuously and through this improve corporate image and gain competitive advantage. In 1985 the concept of EMS was first introduced in Europe (in The Netherlands). The concept is now firmly established in western Europe and is receiving increasing attention in central and eastern Europe, Asia, and South America, as well as in the United States and Canada. EMS (together with environmental auditing) are now widely known and are becoming an integral part of business strategy, and are also being adopted by government as important management tools for companies.

The importance of EMS, both now and in the future, is reflected in the increasing attention being given to it in the international arena.

During the development and implementation of EMS and also during the application of the concepts of environmental auditing and other tools, many companies encountered the need for more clarity on the details of EMS and auditing concepts. Activities on standardization and certification therefore started at both national and international levels. In several countries national standards have been developed, and the ISO is in the process of reviewing and developing standards for EMS, environmental auditing, and several other tools. ISO is a non-governmental organization which was established in 1947 to develop worldwide standards to improve international communication and collaboration, and to facilitate the international exchange of goods and services. ISO is a federation of about 100 national standards bodies representing 95% of the world's industrial production.

---

## 5.6 What Is an Environmental Management System?

An EMS is part of the enterprise's overall management system. It includes the organizational structure, planning activities, responsibilities, practices, procedures, processes, and resources for implementing and maintaining environmental management. It includes those aspects of management that plan, develop, implement, achieve, review, maintain, and improve the enterprise's environmental policy, objectives, and targets.

### 5.6.1 Introduction to Environmental Management Systems

Environmental issues have increasingly important implications for enterprises and other organizations. Depending upon how an enterprise or other organization reacts, environmental concerns can positively or negatively affect the extent to which the organization achieves its goals. The environment presents risks as well as opportunities. Successful enterprises are increasingly trying to manage these risks and opportunities. They do this for at least two main reasons: either to save money, by lowering costs and reducing exposure to liabilities, or to make money, by expanding market share or accessing new markets.

An environmental risk might be contamination of a product to the extent that it is unacceptable to foreign markets, injury or illness of workers or local communities, or a pollution problem that undermines the position of the enterprise in the national or international market.

An environmental opportunity might be the reduction of energy and resource consumption, and therefore costs of production, by reducing pollution or recycling wastes, or it might involve selling the product to a market which imposes environmental requirements. One example of a company that has successfully reduced its energy consumption is the 3M Corporation; through its program with the motto, "Pollution Prevention Pays," its cost savings from pollution control measures have often reached more than US$1 million per year. Enterprises throughout the world are introducing EMS to manage environmental risks and opportunities more systematically and efficiently.

Specifically, an EMS is intended to help your enterprise:

- Identify and control the environmental aspects, impacts, and risks relevant to the organization.

- Develop and achieve its environmental policy, objectives, and targets, including compliance with environmental legislation.

- Define a basic set of principles that guide your organization's approach to its environmental responsibilities in the future.

- Establish short-, medium-, and long-term goals for environmental performance, making sure to balance costs and benefits, for the organization and for its various stakeholders.

- Determine what resources are needed to achieve those goals, assign responsibility for them and commit the necessary resources.

- Define and document specific tasks, responsibilities, authorities, and procedures to ensure that every employee acts in the course of their daily work to help minimize or eliminate the enterprise's negative impact on the environment.

- Communicate these throughout the organization, and train people to effectively fulfill their responsibilities.

- Measure performance against preagreed standards and goals, and modify the approach as necessary.

The integration of environmental management into the overall management function of an organization is critical because the environment is one among a number of external issues affecting the enterprise. A stand-alone EMS would not be effective.

A description of an EMS is: A planned and coordinated set of management actions, operating procedures, documentation, and recordkeeping, implemented by a specific organizational structure with defined responsibilities, accountabilities, and resources, and aimed at the prevention of adverse environmental impacts as well as the promotion of actions and activities that preserve and/or enhance environmental quality.

An EMS follows the well-known quality management approach of "Plan, Do, Check, Improve." It is a problem-identification and problem-solving tool which can be implemented in an organization in many different ways, depending on the precise sector of activity and the needs perceived by management. The specific system implemented depends entirely on the needs and objectives of the organization.

### 5.6.2  Definitions

The terms used are mainly defined with reference to existing models and emerging standards, although these are evolving at the moment and may well be modified (refer to the Glossary for an extensive list of key terms and concepts).

This chapter uses the term "EMS" in the sense of ISO 14001:

> That part of the overall management system which includes organizational structure, planning activities, responsibilities, practices, procedures, processes, and resources for developing, implementing, achieving, reviewing, and maintaining the environmental policy.

This chapter uses the following key concepts and their definitions:

- The "organization" means company, corporation, firm, enterprise, or institution, or part or combination thereof, whether incorporated or not, public or private, that has its own functions and administration. For organizations with more than one operating unit, a single operating unit may be defined as an organization. (ISO 14001)
- The "environment" means, the surroundings in which the organization operates, including air, water, land, natural resources, flora, fauna, humans, and their interrelation. Surroundings in this context extend from within an organization to the global system. (ISO 14001)
- The "environmental effects" that the EMS is intended to manage include any direct or indirect effects on the environment of the activities, products and services of the organization, whether adverse or beneficial. The ISO 14001 standard makes the distinction between: "Aspects," the elements of an organization's activities, products, or services that can interact with the environment, and "impacts," any change to the environment, whether adverse or beneficial, wholly or partially resulting from an organization's activities, products, or services.
- The "environmental vulnerabilities" of an enterprise include risks of injury to workers, the community, and the local or wider environment from the enterprise's activities, products, or wastes, including damage to the functioning and future of the enterprise itself.
- "Continual improvement" means: a process of enhancing the environmental management system to achieve improvements in overall environmental performance in line with the organization's environmental policy. A note is included which clarifies that the process need not take place in all areas of activity simultaneously. (ISO 14001)
- The environmental performance means: measurable results of the EMS, related to an organization's control of its environmental aspects, based on its environmental policy, objectives, and targets.
- Most EMS include procedures for communicating and dealing with stakeholders or interested parties. These are people or other organizations with an interest in the impacts of the organization's activities, products, and services on health, safety or the environment. They can include government regulators and inspectors, investors (including banks and stakeholders), insurance companies, employees, the local community, customers and consumers, nongovernmental organizations, environmental groups, and the general public.

### 5.6.3 The Origin of the Environmental Management System

Enterprises have adopted a range of responses to environmental challenges from doing nothing, to crisis response, to integration of environmental management into the overall management of the enterprise via a well-defined EMS. EMS have benefited from the development of and the experience with two separate management tools over the past 15 years:

- The rising cost of environmental liabilities led companies in North America and Europe to develop environmental auditing as a management tool to identify environmental problems and to monitor the company's environmental performance similar to the way a financial audit is used to measure financial performance. The first goal was to ensure the company's compliance with environmental laws and regulations. Later, the scope was extended to cover the monitoring of "best management practices" for environmental vulnerabilities.

- "Total quality management" (TQM) concepts, although originally aimed at reducing and eventually eliminating defects (noncompliance with specifications) in manufacturing and at improving the efficiency of business processes, have increasingly been applied to managing environmental issues.

Environmental auditing was developed in the 1970s and 1980s by companies such as Allied Signal, Westinghouse, Philips, ICI, and others in response to the mounting cost of not complying with environmental health and safety legislation in highly regulated countries such as the U.S., Canada, and in western Europe. A "compliance audit" monitors compliance with laws and regulations. A "due diligence" or "preacquisition" audit is used by enterprises to identify environmental vulnerabilities and problems of a site or enterprise before any investment is made. The results of the preacquisition audit may affect the level of investment, the purchase price, future investments, and operating costs connected with the site, or even the ultimate decision about the investment.

Today, the term environmental audit has expanded and is sometimes (mis?)used to mean either the Environmental Review to identify an enterprise's environmental impacts and issues, or the EMS. The Environmental Review is performed to identify risks, problems, and environmental opportunities when no EMS is in place; it is similar to the preacquisition audit. The EMS audit is a tool to determine whether the EMS conforms to the organization's planned arrangements and whether it has been properly implemented and maintained.

The origins of TQM derive from the industrial efficiency experts who, during the 1920s, began systematically investigating working methods. The "Deming model" explained below, was developed by Charles Edwards Deming, the "father" of TQM. It has been applied to industrial redevelopment and to environmental management throughout the world.

Historically, enterprises have managed environmental issues three ways:

- *No action.* They do not recognize environmental opportunities or threats until it is too late.

- *Reactively.* Management waits for events and issues to be decided externally before taking steps to deal with them. These enterprises may benefit in the short term, but they will never be certain of approaching problem areas, and are ill-equipped when problems occur.

- *Proactively.* These enterprises monitor environmental issues and concerns as a part of their daily business, and formulate responses before a situation becomes critical. This policy does not mean that these companies escape difficulty, but they are better prepared to deal with crisis situations.

### 5.6.4  Purpose of an Environmental Management System

An environmental review or an audit has at least one serious drawback: although it is a very valuable management tool, it can only describe the environmental situation of the enterprise at the time the review or audit is carried out. The follow-up of nonconformities is not a guarantee that after some time the same or other deficiencies in the organization's control of environmental aspects will not occur. Some form of management system is needed to ensure that the enterprise's environmental targets and objectives are pursued effectively. Furthermore, because relationships between and even within different organizational functional units are often complex, there is a need to have a system of coordination.

An EMS, therefore, is intended to link these different processes through a network of management actions, procedures, documentation, and records, with the aim to:

- Identify and control significant environmental aspects and impacts.
- Identify significant environmental opportunities.
- Identify relevant environmental regulatory requirements.
- Establish a sound environmental policy and basis for environmental management.
- Establish priorities, determine objectives and work toward their achievement.
- Monitor performance and evaluate the effectiveness of the system, including promotion of system improvements and adaptation to meet new and changing conditions and demands.

The communication of the enterprise's environmental objectives and results to external audiences is also gaining importance in environmental management, but is normally not considered a core element in an EMS.

### 5.6.5  Core Elements of an Environmental Management System

EMS will be different for different types of organizations, depending on the nature, size, and complexity of the activities, products, and services, yet all EMS have a number of core elements in common. The core elements include:

- An *environmental policy*, usually published as a written *environmental policy statement*, expressing the commitment of senior management to appropriate environmental management. It is most often understood as a public statement of the intentions and principles of action for the enterprise regarding the environment. The policy statement should define the broad goals the enterprise has decided to achieve. These are most clear if they are quantified.

   For example, a chemical company issued a policy statement: "To reduce emissions of pollutants by 95% within five years." A municipality might adopt the policy: "To provide sewerage and biological treatment of sludge for 60% of the population within three years."
- An *environmental program* or *action plan*, describing the measures the enterprise will take over the coming year(s). The environmental program or action plan translates the environmental policies of the enterprise into objectives and targets and identifies the activities to achieve them, defines employee responsibilities, and commits the necessary human and financial resources for implementation.

In the case of a chemical company, the action plan would identify the steps to be taken by each department to reduce their emissions, commit the necessary funds and staff to meet each goal, and provide for monitoring and coordination of progress toward these separate goals and the overall policy goal. The program also uses the assembled overview of the environmental aspects of the organization and the overview of legal and other requirements that have to be fulfilled. This information is collected for the first time by performing an environmental review.

- *Organizational structures* establishing assignments, delegating authority, and assigning responsibility for actions. In the case of enterprises with multiple sites or different business activities, this includes organizational structures for the enterprise as a whole as well as for the separate operating units. The senior staff member responsible for the environment typically has a direct reporting relationship to the head of the enterprise. Individuals holding strategic or line environmental responsibilities should be adequately supported with human and financial resources.

- The *integration of environmental management into business operations*, includes procedures for incorporating environmental measures into other aspects of the enterprise's operations, such as the protection of workers, purchasing, R&D, product development, mergers and acquisitions, marketing, finance, etc. This includes the development of specific environmental procedures, usually detailed in operating manuals and other operating instructions describing measures and actions to take in the implementation of the environmental program or action plan. Environmental procedures can include:

  - Awareness raising on relevant environmental issues, the environmental policy, the objectives, targets, and the role of every employee in the environmental management system.

  - Internal communication, receiving and responding to communication from external interested parties.

  - EMS documentation and document control.

  - Operational control: procedures and criteria for the operations and activities, as well as goods and services and the suppliers and contractors of the organization.

  - Risk assessment and emergency response plans to identify potential accidents and prevent them from becoming catastrophes.

- *Monitoring, measurement and recordkeeping procedures* to document and monitor the results of specific actions and programs as well as the overall effects of environmental improvements.

- *Corrective and preventive action* to eliminate causes of actual or potential nonconformances to objectives, targets, criteria, and specifications.

- *EMS audits* to check the adequacy and efficacy of the implementation and functioning of the EMS.

- *Management reviews*, the formal evaluation by senior management of the status and adequacy of the EMS in light of changing circumstances.

- *Internal information and training* to ensure that all employees understand why and how to fulfill their environmental responsibilities within the context of their work activities.

- *External communications* and community relations to communicate the enterprise's environmental goals and performance to interested persons outside the enterprise, and to keep them informed about specific environmental issues, difficulties or other matters that may affect them.

EMS and their core elements have been described in several guides and textbooks on EMS but are now also available through the publication of standards. The most important of these is probably the ISO 14001.

The core elements in this standard are structured around the Deming circle. The table of contents of ISO 14001 is as follows:

ISO 14001: Environmental Management Systems — Specification with guidance for use

0   Introduction
1   Scope
2   References
   2.1   Informative references
3   Definitions
4   Environmental Management System
   4.0   General
   4.1   Environmental policy
   4.2   Planning
      4.2.1   Environmental aspects
      4.2.2   Legal and other requirements
      4.2.3   Objectives and targets
      4.2.4   Environmental management program(s)
   4.3   Implementation and operation
      4.3.1   Structure and responsibility
      4.3.2   Training, awareness, and competence
      4.3.3   Communication
      4.3.4   Environmental management system documentation
      4.3.5   Document control
      4.3.6   Operational control
      4.3.7   Emergency preparedness and response
   4.4   Checking and corrective action
      4.4.1   Monitoring and measurement
      4.4.2   Nonconformance, corrective and preventive action
      4.4.3   Records
      4.4.4   Environmental management system audit
   4.5   Management review

## 5.7   How to Develop the Environmental Management System

Few enterprises start from zero when building an EMS. Many, if not most, already have some management procedures or system elements that lend themselves conveniently to the incorporation of environmental issues. Many enterprises have an employee who is responsible for worker protection and health who can also assume environmental responsibilities. The top management of the enterprise, or a plant manager, can take the lead for the development of the EMS. Multinational or major national enterprises are often willing to help their suppliers improve their ability to meet environmental, health and safety, and product standards.

Governments, international organizations such as UNEP and the WHO, universities, business and professional organizations are developing expertise in environmental management and have growing information and training programs.

To provide an overview of how an enterprise can develop and implement an EMS, it is helpful to look again at the Deming Model. The Deming model of quality management, set out in ISO 9000, provides the framework for most EMS. It divides the enterprise's actions into four phases:

- Plan: a *planning* phase — the overall objectives and goals of the enterprise are established, and the methodologies for achieving them are developed.

- Do: an *action* phase — the plan is implemented and the agreed measures are taken in pursuit of the enterprise's goals.

- Check: an *evaluation* phase — the actions taken under the plan are checked for effectiveness and efficiency, and the results are compared to the plan.

- Improve: a *corrective action* phase — any deficiencies or shortcomings are repaired, the plan may be revised and adapted to changed circumstances, and procedures are reinforced or reoriented as necessary.

Virtually every EMS in operation today is comprised of tools derived from this model, and all environmental management standards under development or in the process of implementation follow this model. The new national and international environmental management standards are all based on it (e.g., it is the concept behind the structure of the ISO 14001 standard).

### 5.7.1   Plan:

The enterprise needs to ask the fundamental questions, "Where are we now and where do we want to go?" Answering these questions involves three steps:

- The environmental review: understanding the enterprise's existing environmental position, the requirements put on the enterprise, the relevant environmental aspects, its performance and practices; identifying strengths and weaknesses.

- Obtaining a clear vision of the near future: understanding the likely future of environmental aspects and impacts, and their implications for the enterprise in order to identify risks and opportunities, and an environmental policy: stating how the enterprise will respond to current and anticipated environmental issues.

  Policies must be developed by the senior executives of the enterprise, although proposals may come from all levels of the workforce. Defining goals, however, is a process that must be supported and double-checked throughout the entire chain of command. The lower levels in the hierarchy have an important role to play, as they can ensure that the goals are technically, financially and organizationally feasible. Some enterprises involve external stakeholders, such as government officials, clients, customers, and even environmental groups, in the definition of goals and strategies.

- The enterprise must determine, develop, and implement the necessary structures and procedures and commit the necessary resources to implement the strategy. The development of strategic goals, action plans and procedures build on one another. An environmental action plan must address a number of key issues, including objectives and targets, priorities, responsibilities and accountabilities, schedules and milestones, communications (internal and external), and resource allocation. Again, the more the action plan takes the views and interests of all levels of the enterprise into account, the more realistic it is likely to be and the more understanding and support it will gain from the employees who will be responsible for implementation.

### 5.7.2   Do:

- Responsibilities and procedures should ideally be defined by the people who are charged with their implementation, and then adopted by senior management. Each procedure should be as effective and efficient as possible. The program of training and internal communication should also be defined in the action plan.
- In addition to this, a consideration of the external communication of the enterprise. External communication is an essential element of an EMS, but not all the aspects dealing with external communication have been taken into account in a standard on EMS. Environmental Reporting is a very useful tool, which is still (largely) left to the discretion of the enterprise to be used or to be disregarded.

### 5.7.3   Check:

- The enterprise must have tools to answer the question: "How are we doing?" These monitoring and control tools usually include requirements for records on environmental emissions, wastes, and performance. They also include corrective and preventive actions, environmental audit procedures, and programs. The purpose of this phase is to assess the enterprise's actual environmental performance against its stated policies, and against the objectives and targets in the action plan.

### 5.7.4   Improve:

- A periodic management review will help ensure that the EMS continues to be responsive to changing circumstances, including, for example, new scientific knowledge about the environmental impacts of a chemical, changing national or international markets for products, currency rates, government regulation, and changes in consumer or client requirements.

  The ultimate aim is to continue improving the enterprise's environmental performance. Opportunities for improvement may be found in better implementation of existing programs and policies, technological innovation, new processes and products, new markets, training, etc. An improvement may be as simple as better implementation and control of an existing procedure or the development of a new procedure, or capital investment to correct an unanticipated problem, or even redefinition of the enterprise's environmental goals and objectives in response to internal or external changes in circumstances.

The Deming model is a dynamic one. When the enterprise has identified changes that can or should be made to the EMS, it will inevitably return to the Plan phase to introduce those changes into the environmental policy and action plan.

## 5.8 Enterprise Functions that Will Be Affected

In functional terms, the EMS concerns virtually all business and management activities within the enterprise:

- Research and development: environmental criteria should be considered in product design to meet client demands, regulatory requirements, international standards, or to ensure that products have minimal environmental aspects and impacts throughout their life cycle, from design and raw material use, through manufacturing to distribution, product use, and final disposal.

- Manufacturing: pollution control and cleaner production are obvious issues for the EMS. Other issues may include worker protection, the prevention or mitigation of accidents, and the prevention of long-term, gradual environmental damage from the enterprise's activities or products. Those responsible for manufacturing should be able to count on the EMS to help control the environmental vulnerabilities related to manufacturing processes, including, for example, the selection of appropriate techniques and technologies.

- Finance: finance directors of enterprises in many countries are finding that obtaining financing for projects at favorable rates depends on their ability to demonstrate that their enterprise can control risks, including environmental ones. Moreover, they need to work more closely with the organization's planners to determine the overall financing needs of projects and understand how environmental issues can affect project approvals and the time necessary for receiving these approvals.

- Planning and development: obtaining planning permission for new projects as well as for the expansion of existing operations today often requires completing an environmental impact assessment and making performance guarantees. In many parts of the world, property transactions can result in the acquisition of pollution liabilities from previous activities on the site that must be taken into account in the negotiations.

- Marketing: in numerous countries, consumers have come to expect a certain environmental performance from the products they purchase. Products which have potential severe impacts on the environment may be subject to international regulation or consumer boycotts.

  In Malaysia, for example, the organization Consumers International has organized consumer campaigns that have succeeded in obtaining a ban on a dangerous product in a few weeks. Even if the enterprise does not sell directly to the public in such countries, marketing managers need to understand how these issues can affect their relations with major customers, such as multinationals that do sell to the final customer in such countries and who must meet certain environmental criteria simply to maintain access to these markets. Markets can shift very rapidly once a government has decided to ban or limit a particular product, an intermediate input, or a waste material because of its pollution potential.

- Management and distribution (retail and wholesale): requirements governing packaging and product materials, recovery, and recycling place new demands on distributors in a number of major markets around the globe. The enterprise's EMS needs to take these issues into account, and managers need to ensure that these issues are addressed in the overall management of the enterprise.

## 5.9 Who Will Oversee the System?

An EMS needs the ongoing leadership and support of the highest levels of management in the enterprise if it is going to succeed. In larger enterprises, the full range of issues to be covered by the EMS is entrusted to a high-level environmental management group. This group should be chaired by a senior staff member, such as the president or vice president of the enterprise, or the plant manager. Such a group is often responsible for employee health and safety as well as environmental concerns, since there is a great deal of overlap between these areas.

In some cases, the central group can be quite large, and may have the same policy-making weight as other senior management groups. In other cases, the central group may report to another function, such as manufacturing or engineering, and may support the management of environmental issues in such functional areas.

In other larger enterprises, each operating or functional division might have its own organization and system for the management of environmental issues relevant to that division.

In small or medium-sized enterprises, a small group or even a single person might become responsible for maintaining internal communication with other managers and external communication between the organization and the outside world. This environmental management team or individual would also coordinate activities between divisions and ensure that each external environmental issue that might affect the organization is managed by the appropriate person or division.

## 5.10 Who Should Be Involved?

Since environmental issues can have an impact on virtually every function in an enterprise, an EMS will need to cover all aspects and operations within the organization, sometimes including actions off the site, such as waste or product disposal.

To be effective, the EMS should define and implement the enterprise's policy goals, which are usually decided by senior management. Thus, the EMS depends on the leadership, commitment, and support of senior managers.

The success or failure of the EMS is ultimately decided at the enterprise's operational level. The people charged with implementing its procedures and who are responsible and accountable for environmental performance must be sufficiently motivated and encouraged to carry out their environmental responsibilities with enthusiasm. In order to achieve this commitment throughout the ranks of the enterprise, employees need to be consulted during the development of the EMS. Their comments on what works and what does not will be valuable in the implementation, review, and improvement phases of the EMS. This consultation will encourage all employees to understand and feel a shared responsibility for the success of the system.

## 5.11 Conclusion: Performance and Reward

Performance measures are necessary in an EMS, not only to provide an indication of how well the organization is reaching its goals, but also to serve as an incentive in the pursuit of excellence. Employees need to know that their efforts on behalf of the enterprise are recognized and appreciated. This is best accomplished by rewarding the actions and behavior the enterprise wants to promote. The rewards to the enterprise can also be substantial.

In the short term, even a limited EMS focused on cleaner production can demonstrate that over several years the financial gains can outweigh costs and that inexpensive measures can yield important environmental and financial returns.

Costs may rise in the short to medium term, however, as enterprises invest in the initial environmental review, planning, and implementation of an EMS.

In the longer run, as the ICC *Business Charter for Sustainable Development* and *Agenda 21* state, it is necessary "to recognize environmental management as among the highest corporate priorities and as a key determinant to sustainable development," sustainable development which will secure the growth and prosperity of the current generation of enterprise managers and workers, and of their children.

# 6

## Introduction to Cleaner Production

John Kryger and Rikke Dyndgaard

## CONTENTS

## 6.1 Objective

To explain the basic concepts and definitions of cleaner production.

The reader will gain some understanding of the concepts and practices of cleaner production. The chapter briefly highlights the important issues of cleaner production, and gives some examples and useful data which can be used for presentations or for refreshing your memory on cleaner production.

## 6.2 Introduction

With the passing of time we have come to appreciate more fully the pressure that pollution puts on our natural resources and our health. Europe annually produces 20 million tons of solid waste which is regarded as hazardous. Traditional methods of dealing with such volumes of waste have not been successful, and the resulting contamination of water and land has led to pressure on government and industry to improve the situation.

1-56670-337-9/00/$0 00+$.50
© 2000 by CRC Press LLC

For factory effluents and emissions, the situation is similar. The environmental impacts are increasingly regarded as unacceptable, standards are tightening, disposal costs are increasing. In order to escape this impasse, the authorities and industry are now more seriously trying to find a way to avoid producing wastes and emissions altogether.

This reexamination is happening at a time when increased market competition is in any case forcing companies to make improvements in production efficiency, and generally look for cost-cutting measures. Simple calculations of the market value of chemicals that have been flushed down the drain support the long-standing view of ecologists that emissions, effluents, and other residues, in addition to being pollutants, are in fact wasted resources.

Suddenly cleaner production, pollution prevention, waste minimization, and recycling are more in our day-to-day thoughts. In other words, we are at last thinking more seriously about producing without wastes and emissions, i.e., about "cleaner production." This change in attitude became very noticeable during the United Nations Conference on Environment and Development (UNCED) in 1992. UNCED, in Agenda 21, gives high priority to the introduction of cleaner production methods and preventive and recycling technologies in order to achieve sustainable development. This priority is emphasized in Chapters 20, 22, and 30 of Agenda 21.

This chapter outlines the understanding of cleaner production, what is being done to promote it, and how the concept can be applied in practice.

## 6.3    What Is Cleaner Production?

Cleaner production is a general term that describes a preventive approach to industrial activity. It is neither a legal nor a scientific definition to be dissected, analyzed, or subjected to theoretical disputes. It is a broad term that encompasses what some countries call waste minimization, waste avoidance, pollution prevention, and other similar names, but it also includes something extra.

Cleaner production refers to a mentality of how we produce our goods and services with the minimum environmental impact under present technological and economic limits. It acknowledges that production cannot be perfectly clean. Practical reality ensures that there will be residues of some sort from many processes and obsolete products. However, we can, and must, strive to do better than in the past.

Cleaner production does not deny growth, it merely insists that growth be ecologically sustainable. It is also important that we have a clear view of what cleaner production is not. Some popular misconceptions, for example, that recycling and effluent treatment by themselves constitute cleaner production, have to be refuted, since many vested interests try to repackage existing programs under a new, popular title.

The definition that has been adopted by UNEP is the following:

> Cleaner production is the continuous application of an integrated preventative environmental strategy to processes, products, and services so as to increase efficiency and reduce the risks to humans and the environment.

The main emphasis is clear. It is important to focus on the manufacturing process and also to take a life-cycle approach to products and services themselves.

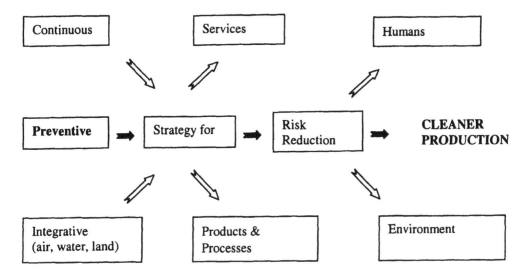

**FIGURE 6.1**
Essential elements of a cleaner production definition.

For **production processes**, cleaner production includes the efficient use of raw materials and energy, the elimination of toxic or dangerous materials, and the reduction of emissions and wastes at the source.

For **products,** the strategy focuses on reducing impacts along the entire life-cycle of the products and services, from design to use and ultimate disposal.

Cleaner production involves applying know-how, improving technologies, and above all, changing attitudes in many places. The essential elements of the cleaner production definition are summarized in Figure 6.1.

### 6.3.1  Why Cleaner Production?

Cleaner production is a good business and environmental proposition to achieve a lower level of pollution and environmental risk. More efficient use of materials and process optimization result in less waste and emissions, which in turn result in lower operating costs. Its focus on occupational health and safety also has positive effects on worker productivity and fewer in accidents. For new processes, such benefits are usually built into the equipment. For older plants, economic incentives can be achieved from process changes or modifications. Examples of benefits of cleaner production are given in Table 6.1.

Cleaner production is especially important to developing countries and countries in transition, because it provides industries in these countries, for the first time, with an opportunity to "leap frog" over older, more established industries which are still saddled with costly pollution control techniques. The following examples show cleaner production applications in developing and in-transition countries.

### 6.3.2  Poland

FSM Sosnowiec manufactures car headlight reflectors, door locks, and window winders. The lamp bodies are made of zinc aluminum alloy, and then copper–nickel–chromium

**TABLE 6.1**

Examples of Cleaner Production

| Industry | Method | Reduction of Wastes and Emissions | Payback Period |
|---|---|---|---|
| Food processing (The Philippines) | Installation of collecting pans to collect fruit drops | Increased productivity: 55 liters of fruit juice an hour | 9 months |
| Automobile component manufacture (Mexico) | Change of floor cleaning detergent to surfactant cleaner, thus increasing coolant recycling system | Eliminated coolant failures, eliminated 4 million liters of oily wastewater per year | 4 months |
| Wood finishing (Malaysia) | Waste segregation | Reduced 54,000 kg of hazardous waste and 5.7 million liters of water | 3 months |
| Cooking works (Poland) | Introduction of a cooler with separate dirty and clean water flows in the benzol recovery plant | Reduced 90% of hydrogen cyanide, toluene, benzene, xylene, and hydrogen sulfide emissions | 1 month |
| Lead oxide manufacture (India) | Replacement of insulation material in furnace | Reduced fuel consumption by 50%, power consumption by 20%, and increased lead oxide production by 3% | Less than 3 months |

Source: UNEP IE, Government strategies for cleaner production, 1994.

plated. The door locks and window winders are made of steel and are zinc-plated. The waste from the factory used to contain cyanide and the following heavy metals: chromium, copper, nickel, and zinc. Then a program to reduce pollution and improve the working conditions at the factory was introduced. New plating processes were introduced, waste from rinses was all but eliminated and a recycling system was introduced, which allowed waste raw materials to be recovered and water reused. This reduced the use of water and raw materials, and waste stream quantities were reduced as follows:

- Chromic acid by 60%
- Copper by 95%
- Cyanide by 80%
- Zinc by 98%
- Wastewater by 93%

wastewater has been purified to the following levels:

- Chromium to 0.1 mg/i
- Copper to 0.1 mg/i
- Nickel to 1.0 mg/i
- Cyanide to 2.0 mg/i
- Zinc to 0.9 mg/i

Capital investments in the program was US$ 36,000 resulting in total savings of US$193,000 per year, with a payback period of two months.

### 6.3.3   Indonesia

PT Semen Cibinong is a cement company near the Indonesian capital of Jakarta. It operates two cement kilns, each producing about 2,000 tons of cement per day. The quality of the cement is determined largely by the firing temperature in the kiln. Too low a temperature, and the cement fails quality standards. Too high a temperature, and fuel is wasted, while production of nitrogen oxides (NO~) and sulfur oxides (SO~) also increases. Both these emissions are dangerous pollutants, the wasted fuel is an unnecessary expense and the low-quality cement is an unsalable product — in other words, an expensive waste of money. The company introduced a sophisticated monitoring and control system to ensure optimum efficiency in the operation of the kiln, particularly in maintaining the right temperature. This system:

- Reduced energy use by 3%
- Increased production capacity by 9%
- Reduced below-standard cement production by 40%
- Produced cement more readily ground to powder
- Leading to further energy savings
- Reduced emissions of NO~

The US$375,000 invested saved US$350,000 per year in energy costs alone and, with the other benefits, paid for itself in less than a year.

The World Bank estimates that 80% of the industrial capacity that Indonesia will be using in year 2010 has yet to be installed. If this 80% adopts cleaner production, Indonesian industries will be able to take full advantage of a rare window of opportunity which for once favors developments in this country over those in the industrialized nations.

### 6.3.4  India

Century Textiles of India is the world's largest exporter of 100% cotton fabrics, and it employs 7,000 people. In 1991–92, its textile division had a turnover of US$99.75 million. Sulfur black is a fabric dye commonly used in India because it does not fade from washing or sunlight, but the traditional dyeing process uses sodium sulfide and produces a foul-smelling effluent containing 30 parts per million (ppm) of highly toxic sulfides. The State Pollution Control Board stipulated that the sulfides be reduced to 2 ppm. By substituting hydrol, a cheap and nontoxic by-product of the maize starch industry, for most of the sodium sulfide, Century Textiles reduced its sulfide emissions to within the 2 ppm limit. The new system produced higher-quality cloth, eliminated the need for new investment to meet the pollution standards, and saved money using the cheaper hydrol. Furthermore, hydrol produces less corrosion than sodium sulfide, meaning lower maintenance costs, and the smell of sulfide in the work place was eliminated (UNEP, 1994).

## 6.4  What Are the Main Barriers to Cleaner Production?

Many companies and countries have shown reluctance to cleaner production. Table 6.2 shows the breakdown of reasons cleaner production is not adopted. The largest obstacle is political, which can stem from resistance to change, misconception, and nonexistent or uncoordinated legislative framework.

Much is said and written about the importance of new technologies. In many instances technology does indeed make a major contribution to preventing pollution, but this is not

**TABLE 6.2**

Reasons Why Cleaner Production Is Not Adopted

| | |
|---|---:|
| Political (60%) | |
|     Bureaucratic resistance | 20 |
|     Human conservatism | 10 |
|     Uncoordinated legislation | 10 |
|     Media sensationalism | 10 |
|     Public ignorance/misinformation | 10 |
| Financial (30%) | |
|     Disposal subsidies | 10 |
|     Scarce money | 10 |
|     Entrenched disposal industry | 10 |
| Technical (10%) | |
|     Lack of centralized reliable information | 5 |
|     Lack of assistance in applying cleaner production to individual needs | 5 |
| Total | 100% |

Source: Huisingh, Don, Extract from U.S. studies, unpublished internal report.

the same as saying that without new technology nothing can be done. It has been shown in various countries that approximately 50% of the pollution generated nationally could be avoided by improvements in operating practices and simple process changes.

Also, many cleaner production programs have reported that once a manufacturer has been *forced* to make process modifications by regulations, the implemented changes resulted in more efficient and lower-cost production lines. Why does, industry wait for government to prompt cost reduction measures? Clearly there is still much to be done before the concepts of cleaner production are universally applied.

## 6.5    Who Is Responsible for Cleaner Production?

Cleaner production means shared responsibilities *and* opportunities. All persons involved in production, distribution, or consumption of industrial products and services carry some responsibility for successful implementation of cleaner production.

In industry, many preventive actions can be taken by individuals during their day-to-day activities. Managers can provide employees incentives to reduce waste and emissions. Plant operators can pay more attention to process optimization, operating practices, and safe chemical handling. Buyers and suppliers can be trained to be conscious of the implications of their decisions on products. Technical personnel can best implement the prevention concept if there is a clear policy framework, and if the assessment methodologies (for example, environmental performance criteria for industrial products) are available, and when information on environmental impacts and alternatives is known. Management support and endorsement of cleaner production is a key factor to all the success.

Government has a crucial role to play, as policies and regulations have significant influence on promoting or inhibiting the development of cleaner production. Because various policies, such as those on industry, trade, finance, environment, and education, can have effects on cleaner production, senior officials need to have a better understanding of how to design and use policies to promote cleaner production. At the same time, government officers involved in permit application, monitoring, compliance, and enforcement need to understand cleaner production options and applications at the plant level to adequately

**FIGURE 6.2**
Cleaner production assessment.

assess and endorse cleaner production. The next section provides additional information on governmental actions.

Industry associations and other nongovernmental organizations can form a strong, collective voice for their members. In addition, these organizations can influence societal values on certain issues.

Increasingly, academia and research institutions are focusing their curricula and research programs on cleaner production, and thus they assist industry in their efforts to contribute to sustainable development. In the long term, educational programs can bring forth a new generation of employers, employees, regulators, and consumers with the cleaner production perspective.

## 6.6   How Do Companies Implement Cleaner Production?

Different tools have been introduced during the last 10 to 15 years to enable enterprises to implement cleaner production. ISO 14001 and EMAS — the environmental management systems — are in focus in the late '90s. An environmental review/audit process is included in these systems, and the potential effect of the environmental management system depends very much on the quality and comprehensiveness of this audit. (An introduction to environmental management systems is given in Chapter 5.)

The development of preventive actions can only occur when the problems are specific and clarified. The cleaner production assessment is used to systematically identify areas of cleaner production so that options for potential preventive actions address the most important sources first. Figure 6.2 shows the consecutive steps of the cleaner production assessment.

The discussion below will elaborate on the generation of options for reducing the wastes and emissions at their source. It should be noted that the same methodology is applicable for the generation of options for reducing risk and reducing energy consumption. As cleaner production is an integrated approach, waste and emissions, reduction in risk, and reduction in energy consumption should always be included as a result of an assessment.

Let us look at these actions in turn. A cleaner production assessment is initiated after a conscious decision has been made by the management to take some action. The first phase is to form the project team, discuss the program with workers and supervisors (who will need to provide much of the data), and document the main processes to be studied (refer to Figure 6.2). It is important to pay attention to psychological aspects of the study. Workers will be reluctant to provide information if they believe they will be punished for process inefficiencies.

From the data provided by plant records and other information, the project team prepares a material balance of raw materials, auxiliaries, waste, energy, products, by-products, and wastes and emissions. When this material balance is of sufficient quality, it will be possible to determine where the main sources of wastes and emissions are. While simple in concept, the necessary information for a good material balance is often difficult to obtain. Many companies do not keep good records of chemicals or discharges.

During the assessment phase, the material balance is studied, and appropriate measures are proposed to reduce or prevent loss of materials.

It is here that the project team uses all means possible to identify cleaner production options. The ideas for options may come from:

- Literature search
- Personal knowledge
- Discussions with suppliers
- Examples in other companies
- Specialized databases
- Some further R&D

A creative intellectual environment to think of all possibilities, based on the widest possible experience, is often needed. Brainstorming and group sessions are regularly used at this stage.

As mentioned above, generating options is a creative process that relies more on inspiration than on logical deduction (although logic remains important). The brainstorming session is a combination of creativity and common sense. Before starting a brainstorming session, literature or other organizations and companies should be consulted, and a site inspection should take place so the generation of options will be more productive. One should focus on all influences of the process that could lead to the generation of wastes and emissions. Brainstorming sessions have proved most effective when managers, engineers, process operators, and other employees, as well as some outside consultants, work together without hierarchical constraints.

It should be noted that during the cleaner production assessment a number of obvious possibilities for immediate improvements may already have been identified. In order to go further, it is often helpful to conceptually divide the process into several essential elements, as shown in Figure 6.3. The option-generating process then considers each element in turn.

- **Change in raw materials.** Changes in raw materials accomplish cleaner production by reducing or eliminating the hazardous materials that enter the production process. Also, changes in input materials can be made to avoid the generation of hazardous wastes within the production process. Input material changes include:
  - Material purification
  - Material substitution
- **Technological change.** Technology changes are oriented toward process and equipment modifications to reduce waste and emissions, preliminary in a production setting. Technology changes can range from minor changes that can be

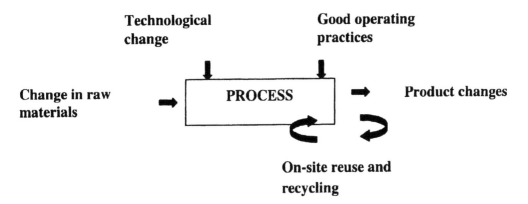

**FIGURE 6.3**
Process elements for cleaner production options.

implemented in a matter of days at low cost, to the replacement of processes involving large capital costs. These include the following:

- Changes in the production process
- Modification of equipment, layout, or piping
- Use of automation
- Changes in process conditions, such as flow rates, temperatures, pressures, and residence times
- **Good operating practices.** Good operating practices, also referred to as good housekeeping practices, imply procedural, administrative, or institutional measures that a company can use to minimize waste and emissions. Many of these measures are used in industry largely as efficiency improvements and good management practices. Good operating practices can often be implemented with little cost. These practices can be implemented in all areas of the plant, including production, maintenance operations, and in raw material and product storage. Good operating practices include the following:
- Management and personnel practices
- Material handling and inventory practices
- Training of employees
- Loss prevention
- Waste segregation
- Cost accounting practices
- Production scheduling

Management and personnel practices include employee training, incentives and bonuses, and other programs that encourage employees to conscientiously strive to reduce wastes and emissions.

Material handling and inventory practices include programs to reduce loss of input materials due to mishandling, expired shelf life of time-sensitive materials, and proper storage conditions.

Loss prevention minimizes wastes and emissions by avoiding leaks from equipment and spills.

Waste segregation practices reduce the volume of hazardous wastes by preventing the mixing of hazardous and nonhazardous wastes.

Cost accounting practices include programs to allocate waste treatment and disposal costs directly to the department or groups that generate wastes and emissions, rather than charging these costs to general company overhead accounts.

By analyzing these factors, the departments or groups that generate the wastes and emissions become more aware of the effects of their treatment and disposal practices, and have a financial incentive to minimize their wastes and emissions. By judicious scheduling of batch production runs, the frequency of equipment cleaning and the resulting wastes and emissions can be reduced.

- **Product changes.** Product changes are performed by the manufacturer of a product with the intention of reducing waste and emissions resulting from a product's use. Product changes include:

  - Changes in quality standards
  - Changes in product composition
  - Product durability
  - Product substitution

  Product changes can lead to changes in design or composition. The new product can thus be made less environmental damaging throughout its life cycle: from raw material extraction to final disposal.

- **On-site reuse and recycling.** Recycling or reuse involves the return of a waste material either to the originating process as a substitute for an input material, or to another process as an input material.

After the options have been generated, an initial selection should be made, considering availability, suitability, environmental effect, and the economic feasibility. This initial selection should take place before the cleaner production option is submitted to more thorough evaluation.

During the feasibility phase, the evaluation will result in a selection of options for implementation. Some examples are given in Figure 6.4.

The implementation phase of the options has to be followed by monitoring of the changes, and inherent to the concept of cleaner production, followed by a new assessment, which will be used to identify new options for cleaner production. This last step closes the chain of continuous improvement.

The procedure outlined above can be used in the same way across the world. We should learn from the positive results that are now coming from those companies and countries that have acquired some experience with the cleaner production process.

Examples of options in each category are given in Figure 6.4.

## 6.7  How Can Governments Introduce Cleaner Production?

The government plays a central role in creating an environment that can expedite cleaner production development and encourage industry to initiate its own cleaner production program.

*Change in Input Material*

| | |
|---|---|
| Printing: | Substitute water-based ink for solvent-based ink |
| Textiles: | Reduce phosphorus in wastewater by reducing use of phosphate-containing chemicals. Use ultraviolet light instead of biocides in cooling tower |
| Electronic components: | Replace water-based film-developing system with a dry system |

*Technological Changes*

| | |
|---|---|
| Filtration and washing: | Use countercurrent washing, and recycle spent wash water |
| Parts cleaning: | Use mechanical cleaning devices; improve parts draining before and after cleaning; use plastic-bead blasting |
| Surface coating: | Use electrostatic spray-coating system; use powder coating systems; use airless air-assisted spray guns |

*Good Housekeeping*

Reduce raw material and product loss due to leaks, spills, drag-out, and off-specification process solution

Schedule production to reduce equipment cleaning. For example, formulate light to dark paints so the vats do not have to be cleaned out between batches

Develop employee training procedures on waste reduction

*Product Change*

| | |
|---|---|
| Batteries: | Replace mercury in batteries |
| Spray cans: | Replace volatile chemicals with water-soluble formulation as aerosol |
| Refrigerators: | Replace CFCs with ammonia |

*On-site Reuse*

| | |
|---|---|
| Printing: | Use a vapor-recovery system to recover solvents |
| Textiles: | Use ultrafiltration system to recover dye stuffs from wastewater |
| Tape measure: | Recover nickel-plating solution using an ion-exchange unit |

**FIGURE 6.4**

Examples of cleaner production techniques. (From H. M. Freeman, Hazardous waste minimization, 1992. With permission.)

The range of tools to catalyze industry to adopt cleaner production is large, and different countries use various combinations to suit their particular needs. In the UNEP IE publication, *Government Strategies and Policies for Cleaner Production*, the available tools are analyzed under four different categories:

- Applying regulations
- Using economic instruments
- Providing support measures
- Obtaining external assistance

In industrialized countries, the first three of these tools have generally been applied in the order given above. Governments have first established regulations designed to limit emissions to the air, water, and onto the land. Economic instruments are then introduced to encourage the observance of these regulations and penalize their infringement. Finally, governments have provided support for industries to enable the regulations to be more easily met. In the process, developed countries have acquired extensive and complicated regulatory systems.

The last tool, obtaining external assistance, is especially relevant to developing countries and those undergoing economic transition. Developing countries may well find it more feasible to depend on raising awareness of the economic benefits implicit in cleaner production. Coupled with suitable support measures and the use of external assistance, it may be sufficient to persuade many industrial leaders to adopt cleaner production procedures, with regulations and economic instruments playing a less important role than they have in the industrialized countries.

### 6.7.1   What Has Been Done in Denmark?

Cleaner production has been actively promoted in Denmark through the Cleaner Technology Programme of the Danish Environmental Protection Agency (EPA). The program was initiated in 1989 as a support program for all types of cleaner technology project activities. The program has gradually developed from specific technical industry support in the early '90s to the LCA and product and consumer focus in the late '90s.

Both Danish EPA and the regional authorities have proactively promoted cleaner production through specific local activities throughout the country.

Denmark may be the leading country in the field of cleaner production, and is definitely the main promoter of the issue through the various support of aid programs supported by Danish EPA and Ministry of Foreign Affairs (Danida).

## 6.8   Other Main Actors in the International Promotion of Cleaner Production

UNEP and UNIDO have established active programs in cleaner production. The programs have the following objectives:

- To increase worldwide awareness of the preventive environmental protection strategy embodied in cleaner production
- To help governments and industry develop cleaner production programs and activities that will expand the adoption of cleaner production approaches and know-how

To meet these objectives, the cleaner production programs carry out the activity elements listed below, which are described in detail in the following sections:

- Technical assistance
- Information dissemination
- Education and training

**Technical assistance.** One of the main focuses of the programs is to promote self-sustaining cleaner production programs in developing countries. Building capacity within an organization, government, or private sector is the most effective way to ensure the implementation of the cleaner production philosophy.

UNEP has joined forces with UNIDO to build capacity in cleaner production by supporting National Cleaner Production Centres (NCPC) in 20 developing countries and countries in transition. The NCPCs are to play a coordinating and catalytic role by conducting demonstration and training programs, assessing national policy and making recommendations, and by acting as a focal point in the country for cleaner production information. The first phase of the NCPC program began in late 1994, with seven centers in operation currently. The joint project is expected to continue for 5 years.

UNIDO and UNEP continue to establish and support NCPCs directly or through other specific project activities. UNEP has recently initiated a project on financing cleaner production investments in developing countries using the NCPC capacity.

**Information dissemination.** The cleaner production programs facilitate the dissemination of information on cleaner production through development of publications, database, and by technical query response.

Publication dissemination allows the program to address a wide audience. The program develops technical manuals, such as a manual on how to conduct waste minimization assessments, and more general publications to raise awareness, such as the popular *Cleaner Production Worldwide* series. Many publications are joint UNEP/UNIDO publications.

The UNEP program has operated an on-line database called the International Cleaner Production Information Clearinghouse (ICPIC), which was originally developed by the U.S. Environmental Protection Agency. Today ICPIC is a disk-based system which contains case studies, publication abstracts, bulletins of events, and contacts. The message center is facilitated on a list serve function on the Internet. UNIDO has a very comprehensive information system on cleaner production available on its web-site.

Both programs have an active query response service, providing direct answers and references by mail and also recently via Internet.

**Education and Training.** Education and training activities are organized, conducted, and/or supported by the cleaner production programs. Each year, the programs participate in a number of workshops, conferences, and seminars to spread the cleaner production message to a variety of audiences.

The UNEP Working Groups of the Cleaner Production Programme are essential in linking the program with expert knowledge. Currently, the working groups are active in the following disciplines:

- Metal finishing
- Textile
- Leather tanning
- Biotechnology
- Sustainable product development
- Policies and strategies
- Education

The working group members provide the necessary technical expertise for training and education activities. In addition, some working groups have organized sector-specific courses and meetings to further spread the cleaner production methodology.

Through these activities conducted in partnership with industry, governments, and other stakeholders, the UNEP Cleaner Production Programme and the UNIDO Sustainable Industry Development Programme continue to play key roles in setting the preventive strategy on the global agenda.

## 6.9   Some Key Learning Points Concerning Cleaner Production

- The cleaner production approach reduces pollutant generation at every stage of the production process in order to minimize or eliminate wastes that need to be treated at the end of the process.
- The terms "pollution prevention," "source reduction" and "waste minimization" are often, in some countries, used to mean "cleaner production."

- Cleaner production can be achieved through good operating practices, process modification, technology changes, raw material substitution, and redesign and/or reformulation of product.

- Effluent treatment, incineration, and even waste recycling outside the production process are not regarded as cleaner production, although they remain necessary activities to achieve low environmental impact.

- The economic advantages of cleaner production are that it is more cost effective than pollution control. The systematic avoidance of waste and pollutants increases process efficiency and improves product quality. Through pollution prevention at the source, the cost of final treatment and disposal is minimized.

- The environmental advantage of cleaner production is that it solves the waste problem at its source. Conventional end-of-pipe treatment often only moves the pollutants from one environmental medium to another.

- The reason cleaner production is slow to be accepted is mostly due to human factors rather than technical factors. The end-of-pipe approach is well known and accepted by industry and engineers. Existing government policies and regulations often favor end-of-pipe solutions. There is a lack of communication between those in charge of production processes and those who manage the wastes that are generated. Managers and workers, who know that the factory is inefficient and wasteful, are not rewarded for suggesting improvements.

- Because cleaner production attacks the problem at several levels at once, introduction of a industry/plant level program requires the commitment of top management and a systematic approach to cleaner production in all aspects of the production process.

These key learning points have been taken from UNIDO's "Training Course: Ecologically Sustainable Industrial Development," Learning Unit 3, 1994.

---

## References

*Government Strategies and Policies for Cleaner Production*, 1994, UNEP (United Nations Environmental Program), Industry and Environment Program Activity Center, Paris.

Huisingh, Don, Extract from U.S. studies, unpublished internal report.

# 7

## A Method in Environmental Management

Hans Schrøder

### CONTENTS

### 7.1  Introduction

This chapter presents a new method called "environmental management by energy-and-materials accounting." The method is based on elementary mathematics and thermodynamics in addition to a principle called "maximum environmental value for your money," that is, optimization of environmental investments. A corporation, or a whole country for that matter, which adopts and applies that principle obtains two advantages: reduced costs and a competitive gain.

We have financial management and we have production management, but we do not, in the same sense of the word, have environmental management. The environmental management that we have is a "soft" and imprecise "science" saying a lot about the problem of setting goals, but little about the art of how to achieve these goals in a cost-effective manner. The results are declarations of intentions, goodwill, and considerable erroneous investments, since today's environmental management is practiced without the foundations on which modern financial and production management are based, namely double-entry bookkeeping and input–output analysis.

Toward the end of the chapter, the thinking developed in the beginning is applied to a well-known environmental problem, namely the emission of nitrogen from agriculture. Step-by-step a properly detailed nitrogen accounting system, a nitrogen input–output model based on elementary mathematics and thermodynamics is developed. This model identifies the reasons for the fact that, in Denmark, one has not yet achieved the goal of a 50% reduction of the nitrogen emission.

After 10 years of work, and at least 15 billion DKK spent, the nitrogen emission from agriculture has been reduced by a mere 15%.

## 7.2    Environmental Management by Energy-and-Materials Accounting*

### Looking Back

In the 1970s we were under the impression that "the solution to pollution was dilution." In the 1980s we thought it was a matter of treating whatever came out of the end of the pipes. In this decade we have realized that neither dilution nor waste treatment suffice.

We are now realizing that we have approached the problem from the wrong end. We have approached it from the output side, not reflecting about why so much smoke and so much wastewater came out of the end of the pipe, but more on how to eliminate both.

Now we face the challenge of turning our attention from the output to the input side and understand the quantities and the qualities of the smoke and the wastewater coming out of the end of the pipes. Before that is done, we cannot really do anything about it, without running an unnecessary risk of wasting resources. We are also slowly coming to terms with the fact that our economy is embedded in the (finite) ecology of the Earth.

In the 1990s, two new buzzwords were coined: *environmental management* and *life cycle analysis* (LCA). And in the middle of the 1990s the concept of *green accounts* (energy and materials balances) was introduced and, in Denmark, actually enforced by law. But the green accounts that have emerged so far have typically been of the soft environmental reporting type with many well-intentioned words rather than energy-and-materials balances.

Environmental management is based on the realization that dilution and waste treatment alone do not suffice. It is necessary to analyze the production systems that generate the pollution, and to manage flows of energy and materials toward goals given in terms of the emission of polluting substances, and at the same time optimize environmental investments. It is a long way from this ideal principle to the reality we see today. We have environmental management systems based on rules and standards, but the environmental dimension remains to be integrated.

A wave of certification schemes overwhelmed corporations in the 1990s. The ISO 14000 family of certifications and the EMAS-registration are among the most important. The certification schemes are based on rules, regulations, and principles but not about "why." There are too many groundless rules, and too little understanding of green accounts and their role in the system's economy.

### Looking Ahead

All enterprises, all corporations seek to maximize their efforts, that is, they seek to get the most out of their resources, regardless of what they do, or are told to do. The better the corporation is at optimization, the better it performs in terms of cost-efficiency and competitive ability.

Perhaps even more important is that corporations understand that the condition for survival and development in a market is that it does *the right things the first time*. It is evident that that rule applies in an economic context, but it applies equally well in an environmental.

Optimization can, of course, only be practiced when it is based on a solid knowledge of how the particular corporation functions. It is not possible to optimize by means of general rules. Therefore, corporations must have greater freedom of action, so they can devote

---

* This section is based on a text in Danish to which Mr. Sten Rønhave, Mr. Lars Vedsø, and Mr. Ole Ravn Jørgensen have contributed significantly.

human resources to carrying out environmental management in an economically optimal way, which, at the same time, must be in accordance with the size of the environmental problem at hand.

One condition necessary for corporations' willingness to engage themselves actively in environmental management is that it pays off. The same holds true for the individual citizen and society at large. In that regard, politicians are charged with the task of creating the incentives for environmental management. But in addition, corporations must have tools which make it possible for them to practice environmental management in a way that is just as rational and goal-oriented as when they practice financial and production management.

At the same time, it must be made possible for environmentally aware consumers (political or green consumers) to make the right decisions with respect to their choice of consumer goods. The possibility for doing so has hitherto been established by means of environmental labeling and life cycle analyses, although it is still only a few products for which consumers can obtain relevant information and hence have a real option. Also, in this context, a series of standards and schemes exist, but they are mutually competitive and thus tend to create confusion rather than clarification.

An alternative way of approaching the problem is to set up the energy and materials balances at a suitable level of detail. This is central to the method *environmental management by energy-and-materials accounting*. It is an easily understood and scientifically foolproof tool made for setting up energy-and-materials balances and practicing environmental management based hereon. Furthermore, the basic tool: input–output analysis, is well-known in financial and production management. Hence, the method has the potential of placing environmental management as a discipline in line with financial and production management. That is the long-term challenge. The short-term challenge is to get green accounts organized to make it possible to actually utilize them as a powerful tool in environmental management.

### What Is Environmental Management? And What Is Green Accounting?

The words *environmental management* and *green accounting* have come to stay. But what are we talking about? What is environmental management? What is management? And what is green accounting?

No industrial manager would consider managing a corporation without financial accounts. Accounting, calculations, budgets, analyses, etc., all belong to the backbone of management.

Green accounts are accounts of energy, water, nitrogen, carbon, phosphorus, and so on through the list of elements. Financial accounts are for economic management what green accounting is for environmental management: they are necessary (albeit not sufficient) conditions for management.

There is a lot of talk about environmental management, and a lot of money is spent. Think of the expenses incurred by the two comprehensive Danish environmental management projects, the Aquatic Action Plan, and the plan to reduce the emission of $CO_2$ from Denmark (the Energy Plan). Add to that, the money that has been spent by the private sector over the past five years. The total sum probably adds up to a three-digit billion amount in DKK. Society and its corporations could save large amounts of money by rational environmental management by applying energy-and-materials accounts in their effort to attain the goal *maximum environmental value for your money*. They can do this using energy-and-materials accounting.

### From Economic and Production Management to Environmental Management

Environmental management is not management of the environment. Environmental management is management of energy and materials (the metabolism) flowing through industries, corporations, institutions, and households. Environmental management has a sister in financial management, which can be seen as the management of money flowing through the corporation. Economic management has a sister in production management, which is management of the goods and services that flow through the corporation.

Having said *economic management* one also has to say *production management* since there is a direct link between the exchange of money in financial management on the one hand, and the exchange of goods and services on the other.

Production management is thus management of the flows of goods and services since these flows are linked to the flows of money. Production management and economic management are two sides of the same coin, since the two sets of flows are reflections of each other.

There are not only these two, but three sets of flows. In addition to the mutually connected flows of money and goods and services, we have flows of energy and materials. As sure as the flows of money are interconnected with the flows of goods and services, just as sure is it that we have energy and materials flows that also, in an unambiguous way, are connected with the first mentioned sets of flows. Hence, we have not just two, but three sets of interconnected flows.

### A Bit of Accounting Philosophy

The accounting matrix consists of nine elements (see Table 7.1). The matrix has three rows. Read from below and upward it is the keeping of accounts, then the presentation, and finally the analysis of accounts. The three columns are, read from left to right, economy, production, and ecology. The columns are the "dimensions" of the account.

Nowadays the term *accounting* has many connotations. In addition to financial accounts, we have ethical accounts, social accounts, and knowledge accounts, but they are not accounts in the real sense of the word, because they have no common currency and cannot therefore be added. They concern not only something material and factual, but in addition, human values. They are the so-called soft accounts in which the term *account* is not used in the usual sense of the word, namely to "account for," but in the sense to "call somebody to account." The hard accounts, on the contrary, concern material or factual matters, in other words, things that can be recorded and measured in physical units, like DKK, man-days, kg, $m^3$, and Joule.

Unfortunately, the common conception is that green accounts belong in the soft category. That, however, is a mistake. Green accounts naturally belong in the hard category, where they are safely anchored in the environmental dimension, although they have not yet been integrated into accounting and management systems.

As stressed above, the accounts are interconnected. Thus, if one of the entries in one of the accounts changes, then the two other accounts also change. This is just mentioned in

**TABLE 7.1**

The Accounting Matrix and the "Hard" Accounts

|  | Economy<br>Money | Production<br>Goods and services | Ecology<br>Energy and materials |
|---|---|---|---|
| Analysis of accounts | Financial management | Production management | Environmental management |
| Presentation of accounts | Financial accounts | Production accounts | Green accounts |
| Keeping of accounts | Bookkeeping | Production databases | Green bookkeeping |

passing since the remainder of this chapter is not about the interconnection between the three dimensions of the account, but only about the last, but not yet integrated dimension, the environmental dimension.

### The Objective and the Task of Environmental Management

What is the objective of environmental management? The goal is determined politically as a result of the policy of the corporation. It is a political task to set the goals for the direction in which the corporation or the society should be headed. It is a technical task to manage and steer toward these goals and to attain them quickly and cost-effectively. That applies to environmental management of a corporation, as well as to society at large. The ultimate task is to obtain maximum environmental value for the money, that is, to optimize environmental investments.

Since flows of money and flows of goods and services are linked to flows of energy and materials, it is evident that our management tools are incomplete. They do not include the environmental dimension. They do not account for flows of energy and materials, that is, the green accounts that underlie any economic account.

The WCED report (the Brundtland Report, 1987) says that the global task is to attain "sustainable development." The report is rather vague with respect to the goal, but clear in its answer to the question of how to attain it, namely: by "producing more with less," that is, to produce more goods and services with smaller inputs of energy, water, nitrogen, carbon, etc. That is equivalent to saying that the production efficiency in terms of energy and materials must be increased.

What does the environmental dimension look like? How is it possible to record and account for energy-and-materials flows through a system that can be as large as the whole of society, or a trade, agriculture for example, or as small as a factory or a household? And when that is done, how can one then apply it as a management tool?

### Recording of Flows and Accounting Are Inseparable

It is important to establish what one's data are intended for before embarking on a scheme to record them. If that is not done, then one works according to the OOPS method: OOPS, that one did not do us any good. Or, OOPS that one we really did not need, and so on. The art of the matter is to make the accounting visible and easy to understand so that it can be used as the basis of the planning, completion, and analysis of the measurements.

One can draw the flows of energy and materials as they pass through the corporation. That corresponds to establishing a plan of accounts in a double-entry bookkeeping system. That plan is the backbone economic accounting. In the same way, any green account has a backbone, a plan of accounts, that can be visualized by drawing it as a system of boxes (accounts) and arrows (flows).

When the environmental account plan is established and drawn one can start the recording because now one knows precisely for what purpose the recordings are made. We can point at the drawing and say: that one and that one must be recorded with such and such an accuracy and detail.

### Analysis of Green Accounts

When the first green account is established, one can analyze it as one analyzes a common financial account. One of the tools in analysis of financial accounts is to form key figures, that is, ratios of flows. They are dimensionless numbers given in percentage of something of the same currency.

In the analysis of the green accounts, one can form similar ratios. They are typically efficiencies, and they are key numbers of the same kind as in analysis of financial accounts, but they deviate from them on one important point, namely: that green key numbers in general (so-called *coefficients of performance*), and the efficiency in particular, are mathematically well-defined quantities subject to the laws of nature.

The green key figures, the coefficients of performance, are elements in the structure of the system and can be defined to a point at which we can go backwards. Hence, whereas we began by forming green key numbers from the flows, we can now calculate flows from key numbers and inputs. Then we have a model, an input–output model that can be applied for environmental management.

### From Objectives to Maximum Environmental Value for Your Money

When the objective of the environmental management is formulated, the remainder is a technical task, namely to determine how we obtain the maximum environmental value for our money, that is, maximum nitrogen or $CO_2$-reduction for the investments made to reduce the emission of these substances. We establish that our green accounts are such and such, and we wish they should be such and such in a matter of five years, for example. How do we attain that in the most cost-effective manner, that is, the manner that requires a minimum input of goods and services? It is the task of environmental management to find the answer to that question. Energy and materials balances are necessary, albeit not sufficient.

Think of a computer game, a game of chess, for example. That game is played according to certain rules invented by people. In the same way, one can imagine a computer game played according to the laws of nature and the rules of mathematics. The previously mentioned input–output model is a green accounting game that observes the laws of nature, just like a chess game on the computer observes the rules of chess. That is to say, it is no longer a game, but a tool for environmental management. Whenever the player "touches" the key numbers, he or she gets a new account in which the rules of the game are strictly observed.

### 7.2.1  A Method in Environmental Management

#### A Bit of Thermodynamics

The history of green accounting goes all the way back to 1824 when Carnot, a French engineer, wrote a memoir on the efficiency of heat engines. He showed that we must have two reservoirs in order for a heat engine to work: a hot (the boiler) and a cold (the environment). The energy "falls" from the hot to the cold reservoir just like the water in a hydropower station falls from the reservoir to the downstream river. It is out of that flow of energy that we can derive, or extract, some mechanical energy, but how much? This is what Carnot wanted to find out. He depicted the situation as sketched in Figure 7.1.

He defined the efficiency as the derived mechanical energy in proportion to the throughput and found that the ideal, the maximum, efficiency is:

*Equation 1    The Ideal Efficiency of a Heat Engine (Figure 7.1)*

$$a = \frac{T_0 - T_w}{T_0} = 1 - \frac{T_w}{T_0}$$

in which $T_0$ is the temperature of the boiler and $T_w$ is the temperature of the environment. It appears that the efficiency is always smaller than one, which reflects the fact that no processes in practice are reversible. They cannot, by themselves, run backward.

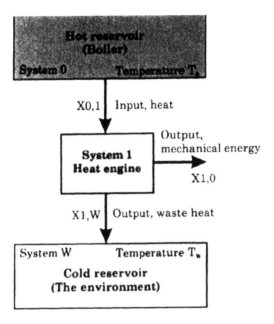

**FIGURE 7.1**
Carnot's perception. Energy only.

The input–output model, which describes the system in Figure 7.1, is straightforward. The two outputs are calculated by two simultaneous conditions (equations). The first condition is that the output of energy equals the input:

*Equation 2   Output Equals Input in Steady State (Figure 7.2)*

$$X1,0 + X1, W = X0,1$$

The energy flows *through* in other words. The other equation defines the efficiency:

*Equation 3   Definition of the Efficiency (Figure 7.2)*

$$a = \frac{X1,0}{X0,1}$$

The model solves the two equations, that is, it calculates the two flows as a function of the management options that exist, namely to change the input (X0, 1) or to change the efficiency (a) or both.

### Generalization of Carnot's Perception

Carnot's perception is not only the perception of how a heat engine works, it is the perception of how any production system works. Just as we must have two reservoirs for a heat engine to be able to perform work, we must have two reservoirs to drive any production system, namely a resource and a waste reservoir. Out of the flow from the resource to the waste reservoir we can extract something of utility for us, namely food or industrial products, including energy (see Figure 7.2). But the result is evidently a degradation of the quality of the resources.

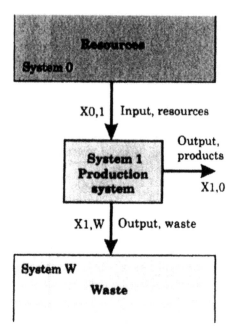

**FIGURE 7.2**
The generalized perception. Energy and materials.

The systems in Figure 7.1 and Figure 7.2 are analogous. Figure 7.2 is an extension of Carnot's consideration, a generalization so that the accounting no longer concerns just energy, but now also materials.

### The Software System

A computer system, SteadyWin (located at www.danedi.com), has been developed by Danedi with the purpose of accounting for energy and materials in systems like the ones shown in Figures 7.1 and 7.2. It is a system with which one can construct all conceivable green account plans. It is in reality an infinity of green accounts and input–output models with mathematically defined management options.

By means of the building blocks in the figures one can construct green accounts and input–output models that describe systems as large as a country, or a county, or as small as an industry or even a household. The size and nature makes no difference. It can be a green account for an industrial complex or a forest lake, a man-made or a natural ecosystem.

In later years a new term *industrial ecology*, has emerged, and one speaks about the metabolism of a society or an industry. They are good words because they guide us to think of the fact that a corporation has an ecology, that is, it has a throughput of energy and materials, just like natural ecosystems.

### Detailing the Plan of Accounts

Development of green accounts and input–output models require first that one agrees to do so and second, that one has acquired some practice in doing it. Under all circumstances the method is to begin with the basic account, as shown in Figure 7.2

When the basic account has been set up, one must detail it, i.e., understand the single system as the result of several subsystems in interaction. The system in Figure 7.2 can thus, for example, be understood as a result of an interaction between four subsystems as shown in Figure 7.3.

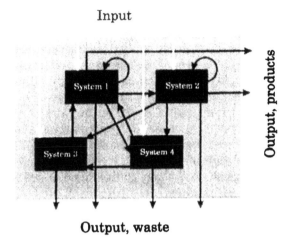

FIGURE 7.3
The system in Figure 7.2 understood as four subsystems in interaction.

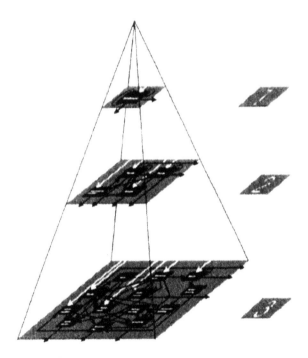

FIGURE 7.4
The pyramid of detail, the top-down approach.

The line of thought that brought us from Figure 7.2 to Figure 7.3 can be reused on subsystems' subsystems until, in the end, one counts atoms. But then one has gone too far since the art of the matter is to stop at a level of detail which is complex enough to be realistic, but yet so simple that everyone can understand it. One can imagine the detailing process as a pyramid or a set of Chinese boxes. When we zoom in all the way before the first box is lifted, we are at the top of the pyramid. When we zoom out all the way, we are in the landscape itself. The challenge is to find a level of detail, that can serve as a map of that landscape (see Figure 7.4).

### Life Cycle Analysis, Environmental Management, and Environmental Declarations

Life cycle analysis (LCA) is one of the most important tools used today in environmental management. At the same time it is the only applied tool that can be said to have a scientific basis. The purpose of using LCA is to evaluate a product's effect on the environment. The method has three obvious shortcomings. First, it is extremely difficult to delimit the task. Second, the method is complex. Third, it yields ambiguous results.

For many years to come, carrying out LCA will be the work of specialists, and in addition, LCA is costly. It is therefore evident that LCA cannot stand alone. It has to be supplemented with energy and materials balances, green accounts. In LCA one traces a product from cradle to grave, from the mine to the landfill. This is not the case in the method of environmental management by green accounting. In that method one looks at the entire economic system that produced the product.

LCA and the environmental management by green accounting methods have a common goal, namely to quantify the environmental effect of industrial products. Some time in the future it is envisaged that it will be possible to label products with a declaration that not only informs about the content of the product, but also about what it has taken in terms of energy consumption and emissions of polluting substances to produce, use, and deposit it after use.

In the effort to formulate unambiguous environmental declarations, LCA is to be considered complementary to the method of environmental management, since neither of the two methods by itself can pinpoint the scientific truth about the environmental effect of an industrial product from cradle to grave. However, it is our considered opinion that the method of environmental management by green accounting will gain momentum, not at the expense of, but as a supplement to LCA since it is superior on the three points mentioned above: (1) the task can be delimited, (2) the method is easy to apply, and (3) it yields unambiguous results.

### Why, How, Who, and When?

A managing director or an environmental manager is entitled to raise the question: Why should we apply the method of environmental management by green accounting on top of the tools and tasks we already have to cope with? The answer is that environmental management by green accounting should be used for the same reasons that the corporation uses money accounting in its financial management and input–output analysis in production management. One can hardly imagine financial management and production management without these tools, just as one cannot navigate a ship without modern navigation equipment. It simply does not make sense to carry out environmental management without accounting for energy and materials. Words do not suffice. First, the new method can be used to reduce costs (maximum environmental value for your money), and second, to acquire a competitive lead in the market.

Holistic or lateral thinking has been the trend in modern management for some time. With the attempt to link economy, production, and environment, a real alternative exists to bring management tools together into a single system in which the combination of recordings, registrations, and reporting becomes a valuable basis for corporations' decisions.

### Conclusion

Environmental management without green accounts (energy and materials accounts) does not make sense. The question is not *whether* it is necessary to expand the accounting by integrating the environmental dimension. The question is *when* it will be done. When it is done, the road is paved for optimization of investments in environmental projects. Economically as well as environmentally it will make a considerable difference. In fact, it is no exaggeration to say that it will be a revolution.

## 7.3   Application

### 7.3.1   Environmental Input–Output Analysis

The analysis begins by examining flows of energy or materials across a single system's boundaries. It proceeds in what Odum (1989) calls a "top-down" or "outside-to-inside" approach by considering this system as interacting subsystems.

The number of interacting subsystems is initially small, and is increased in a stepwise procedure until there is harmony between model complexity, on the one hand, and quality and availability of data on the other. A model resulting from the analysis should be complex enough to be realistic, yet simple enough to understand.

### *First Level*

Consider an agricultural system as a one system, as shown in Figure 7.5. Inputs of new nitrogen may either be recovered as protein in net exported products (P), lost as waste nitrogen (W), emitted to the atmosphere or the hydrosphere.

In addition to flows there are funds to consider. It is necessary to distinguish between these two simple and distinct categories, especially in analysis of agricultural systems. The *flow* category includes the elements that figure in the account *only* as inputs or outputs. The *fund* category comprises elements that appear *both* as inputs and outputs: the elements that enter and come out of the process in the same amount (Georgescu-Roegen, 1976). As an example, if a farmer sows one bag of barley and harvests 40, then one bag is fund and 39 are flow.

Inputs enter through the top of the control volume, the rectangle in Figure 7.5. Nitrogen in products exits the right-hand side, whether they return to the system as a fund (F) or not (P). Wastes (W) exit the bottom in an amount equal to I–P in steady state.

The first law of thermodynamics, and the steady-state assumption, imply that input equals output, see the left-hand side of Figure 7.5:

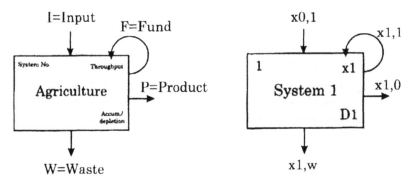

**FIGURE 7.5**
Single system. First level.

*Equation 4   Input Equals Output (Figure 7.5)*

$$I = P+W$$

The problem has three unknowns, namely the three outputs P, F, and W. Therefore, it remains to establish two additional equations to close the system of equations. One could do that by normalizing the unknowns, less one, by their sum, the system's throughput. Then the COPs are two *output coefficient* equations which close the system of equations:

*Equation 5   First Output Coefficient (Figure 7.5)*

$$^a\!P = \frac{P}{P+F+W}$$

and

*Equation 6   Second Output Coefficient (Figure 7.5)*

$$^a\!F = \frac{F}{P+F+W}$$

The software system allows one to define the COPs in other and physically meaningful ways. For instance, the efficiency is defined as (see Figure 7.5):

*Equation 7   Definition of the Efficiency (Figure 7.5)*

$$a1,0 = \frac{P+F}{I+F}$$

which replaces Equation 5.

The last equation is a product allocation equation, a reciprocal fold equation. If F is the fund of seeds returned to the field, out of the total harvest P+F, the farmer speaks of a (P+F)/F fold increase on his seed fund. I use the reciprocal:

*Equation 8   The Reciprocal of the Fold Increase (Figure 7.5)*

$$a1,1 = \frac{F}{P+F}$$

to obtain the closing equation and replace Equation 6.

COPs are dimensionless numbers in the range between 0 and 1. They are ratios of linear combinations of known or unknown flows and funds.

The two systems of equations:

                    A. Equations 4, 5, and 6

and

                    B. Equations 4, 7, and 8

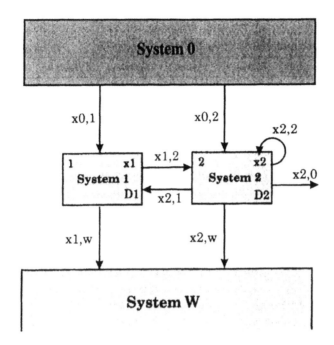

**FIGURE 7.6**
Two systems in interaction. Second level.

are both valid equation systems that determine the three unknowns, the outputs, from the system in Figure 7.5 as functions of the input and two COPs, formed either as output coefficients, or user-defined COPs.

Choosing option B, the three governing equations in matrix form are:

*Equation 9   Three Equations Ready to be Solved (Figure 7.5)*

$$
\begin{bmatrix}
0 & 1 & 1 \\
(1-a1,0) & 1 & 0 \\
(1-a1,1) & -a1,1 & 0
\end{bmatrix}
*
\begin{bmatrix}
F \\
P \\
W
\end{bmatrix}
=
\begin{bmatrix}
I \\
a1,0*I \\
0
\end{bmatrix}
$$

The graph in Figure 7.5 is a directed graph. Flows are positive in the direction indicated by the arrows and negative in the opposite direction. Hence, flows are positive by definition. For instance, if W were negative, it would imply that we were able to convert waste into useful products, or that heat, by itself, can flow from cold to hot reservoirs in violation of the second law of thermodynamics. The condition that all flows (and funds) be nonnegative is therefore a necessary, albeit not sufficient condition, to meet that law.

### Second Level

At the first level of analysis, the three outputs are functions of the input and two COPs of the main system. Stepping to the second level, one can likewise determine outputs from each subsystem as functions of their inputs and COPs.

Consider a production system consisting of two systems interacting as indicated by the directed graph in Figure 7.6.

**TABLE 7.2**

The Input–Output Matrix of the System in Figure 7.6

| From | Into | System 0 | System 1 | System 2 | System w | Total Output |
|------|------|----------|----------|----------|----------|--------------|
| System 0 | — | | x0,1 | x0,2 | — | — |
| System 1 | 0 | | 0 | x1,2 | x1,w | x1 |
| System 2 | x2,0 | | x2,1 | x2,2 | x2,w | x2 |
| System w | — | | — | — | — | — |
| Total input | | | x1 | x2 | | |

Subsystem 1 in Figure 7.6 has two outputs, while subsystem 2 has four. Hence, we have a total of six subsystem outputs, and hence six unknowns. Since we have two balance equations, we must define four COPs to close the system of equations. One of these comes from subsystem 1 (which has two outputs), and three from subsystem 2 (which has four outputs).

The number of COPs required to close the system of equations related to a subsystem, is equal to the number of outputs from that subsystem less one. To clarify this, imagine that the system in Figure 7.6 is a network of water pipes joined at two nodes, the subsystems. Water arriving at the first node is distributed out of its two outlets by one adjustable valve. The balance equation makes sure that the remainder leaves through the other outlet.

The second node, with four outlets, requires three valves. The total number of valves (COPs) necessary and sufficient to control all flows and funds is one in subsystem 1 and three in subsystem 2.

The directed graph in Figure 7.6 translates into the input–output matrix in Table 7.2.

In general terms, the symbol xi,j, which at times is also written as $x_{i,j}$, stands for the flow of energy or matter *from* system i *into* system j; $x_{j,i}$ stands for the flow *from* system j *into* system i; and xi,i is the fund, the return flow from subsystem i back into system i.

All *inputs* into system j therefore have j as the *second* index on x, and the total input to, and output from, system j is $x_j$. All *outputs* from system j have j as the *first* index on x.

In input–output economics, Leontief (1966) defined *input coefficients* as the *quantity of the output of sector i absorbed by sector j per unit of its total output j.* The *input coefficient* of product of sector i into sector j is:

*Equation 10   Input Coefficients*

$$a_{i,j} = \frac{x_{i,j}}{x_j}$$

For example, $a_{1,2} = x_{1,2}/x_2$ is the input *into* system 2 coming from system 1 *per unit of the total output from* system 2.

The present environmental input–output analysis applies *output coefficients* defined by Ayres (1978) as the fraction of outputs from system j that go into system i:

*Equation 11   Output Coefficients*

$$a_{j,i} = \frac{x_{j,i}}{x_j}$$

For example, $a_{1,2} = x_{1,2}/x_1$ is the output from system 1 going into system 2 — per unit of the total output from system 1. Output coefficients are default COPs in the software system.

**TABLE 7.3**

The Structural Matrix of Output Coefficients for the System in Figure 7.6

| From | Into | System 0 | System 1 | System 2 |
|------|------|----------|----------|----------|
| System 0 | | — | — | — |
| System 1 | | a1,0=0 | a1,1=0 | a1,2=x1,2/x1 |
| System 2 | | a2,0=w2,0/x2 | a2,1=x2,1/x2 | a2,2=x2,2/x2 |

Reading horizontally in Table 7.3, it is seen that a complete set of output coefficients for the system in Figure 7.6 are the four nonzero elements of the structural matrix.

These COPs can be thought of as adjustable valves in the hydraulic network metaphor. They can be positioned at any value in the range 0 to 1, excluding combinations resulting in negative flows.

COP equations that are not efficiency equations are called *product allocation equations*. A full exposition of the equations governing the flows and funds in Figure 7.6 is given below:

### Subsystem 1

*Equation 12   Input equals output*

$$x0,1 + x2,1 = x1,2 + x1,w$$

*Equation 13   Efficiency*

$$a1,2 = \frac{x1,2}{x1,2 + x1,w}$$

### Subsystem 2

*Equation 14   Input equals output*

$$x0,2 + x1,2 + x2,2 = x2,0 + x2,1 + x2,2 + x2,w$$

*Equation 15   Efficiency*

$$a2,0 = \frac{x2,0 + x2,1 + x2,2}{x2,0 + x2,1 + x2,2 + x2,w}$$

*Equation 16   Product allocation*

$$a2,1 = \frac{x2,1}{x2,0 + x2,1}$$

*Equation 17   Product allocation*

$$a2,2 = \frac{x2,2}{x2,0 + x2,1 + x2,2}$$

In matrix form we get:

**Equation 18**    *Six Equations Ready to be Solved after Specification of the COPs ($a_{i,j}$) and the Inputs $x0,1$ and $x0,2$ (Figure 7.6)*

$$
\begin{bmatrix}
1 & 1 & 0 & -1 & 0 & 0 \\
(1-a1,2) & -a1,2 & 0 & 0 & 0 & 0 \\
-1 & 0 & 1 & 1 & 0 & 1 \\
0 & 0 & (1-a2,0) & (1-a2,0) & (1-a2,0) & -a2,0 \\
0 & 0 & -a2,1 & (1-a2,1) & 0 & 0 \\
0 & 0 & -a2,2 & -a2,2 & (1-a2,2) & 0
\end{bmatrix}
*
\begin{bmatrix}
x1,2 \\
x1,w \\
x2,0 \\
x2,1 \\
x2,2 \\
x2,w
\end{bmatrix}
=
\begin{bmatrix}
x0,1 \\
0 \\
x0,2 \\
0 \\
0 \\
0
\end{bmatrix}
$$

from which the six unknowns can be determined as function of two inputs (the vector on the right-hand side), and four COPs (two efficiency and two product allocation COPs). In general, if n is the number of subsystems, then the number of unknowns is k+n, which are functions of k COPs in addition to inputs.

In this analysis, the COPs are perceived, not as coefficients, but as management options.

## Higher Levels

One can continue to identify systems within systems to obtain more detailed hierarchical structures, longer vectors of inputs, and larger system matrices. But contrary to input–output economics, data available in environmental input–output analysis seldom warrants a very high level of analysis.

Stepping to higher levels of analysis, one simply repeats the exercise on each subsystem, recalling that they themselves are nothing but interacting sub-subsystems. Any resulting structure, no matter how large and complicated, is thus built by repetitive use of simple principles.

Suppose a system can be described by n subsystems in interaction. From each subsystem there are two possible outlets to the exterior (products and waste) in addition to n outlets to interior subsystems. The highest possible number of unknowns, and therefore equations necessary to describe the network, is $N = n(n+2)$.

At first level, n = 1 and hence N = 3. At second level, n = 2 and N = 8. Suppose n = 10, then N = 120 equations. This would apply in case the matrix were full of nonzero elements. In environmental input–output analysis, however, the input–output matrix is significantly "diluted" by zeros. Hence, the number of equations necessary to close the system is much smaller than the largest possible number.

## The Software System SteadyWin

With enough time and patience, one could in each case work out the mathematics, that is, write the equations, convert them to matrix form and solve them. A software system, SteadyWin,* has been developed by Danedi to do that. One draws the system network on the computer screen. Then one defines necessary and sufficient COPs. The software provides a mathematical mapping of flows and funds as functions of the management options: inputs and COPs.

One can change the management options in preselected numerical steps at a rate limited only by the speed at which the computer solves the equations and pastes solutions onto the directed graph on the screen.

---

* A demo version of the software, including the files used in this chapter, is available on www.danedi.com

### 7.3.2 Accounting for Nitrogen in Danish Agriculture

The following deals with the subject of managing nitrogen in agriculture in general, and in Danish agriculture in particular. Such analysis is urgently needed because the nitrogen emission issue has reached the agenda of policy-makers due to rising levels of nitrate in groundwater and oxygen depletions in the open sea.

Denmark pioneered the now generally accepted notion that countries bordering the North Sea and the Baltic Sea should reduce their nitrogen emissions by 50%. In Denmark, less than 10% of the nitrogen emission comes from nonagricultural sources. The 50% goal therefore applies to the emission of nitrogen from the agricultural sector. In 1987, Denmark launched an ambitious plan to achieve this goal.

In the following, the software system SteadyWin is applied to set up the nitrogen account (nitrogen balance) for Danish agriculture in the middle of the 1980s and to compare it to the balance in the middle of the 1990s.

### *First Level*

Before considering the nitrogen balance for agriculture, one must delimit the control volume. Think of a single farm, or a group of farms, contained in a box with a lid, vertical sides, and a bottom. Assuming that the stock of nitrogen inside the control volume is constant, say as an average over three years, we know that nitrogen entering the system, regardless of form, sooner or later leaves the system in one form or another.

The lid of the control volume can be thought of in terms of the contours of the landscape, including buildings and machinery, displaced, say a couple of meters upward. The bottom can be imagined as an interface displaced a meter below ground level to include the root zone. The sides are an imaginary projected fence with gates, through which nitrogen is exchanged with other sectors of the economy.

It is the purpose of agriculture to produce plant and animal products. Therefore, it is appropriate to initiate the analysis by considering what is produced by the plant and animal production systems and how it is utilized.

Consider the nitrogen flows and funds in Danish agriculture averaged over the period 1983/84 to 1986/87. The agricultural area at that time was 2,863,000 hectares. The *unit* is $10^6$ (M) kg nitrogen per annum.

In Table 7.4, only three entries belong to the flow category, namely export of plant products (75 units), export of animal products (81 units), and burned straw (7 units). All other entries belong to the fund category.

The funds identified so far are what one might call *intended* funds. On the waste side we have an *unintended fund*, namely the volatilized ammonia–nitrogen, which is subsequently redeposited through the lid of the control volume. Although unintended, it is evidently a fund, see Figure 7.7.

In the mid 1980s, Danish agriculture received about 49 units of ammonia–nitrogen in wet and dry deposition, of which 29 were estimated to come from sources within the control volume. These 29 units are the unintended fund, and the remaining 20 are a flow, an input of new nitrogen coming from sources external to outside Danish agriculture.

Having identified flows and funds, one can complete the balance at the first level of analysis with a proper distinction between flows and funds as shown in Table 7.5 where entries transferred from Table 7.4 are boxed.

The entry column in Table 7.5 is summarized by the graph shown in Figure 7.7. It reads as follows: at an input of 635 units of new nitrogen, an input of 295 units of product funds, and 29 units of unintended fund, the system produces funds in the same amount in addition to a product output of 156 units and a waste output of 479 units.

**TABLE 7.4**

Agricultural Products and Their Utilization
in Denmark (in million kg N per annum
over a three-year period in the mid 1980s)

| Plant products and their utilization | 363 | |
|---|---|---|
| Domestic fodder and bedding | 268 | Fund |
| Returned | 6 | Fund |
| Export of plant products | 75 | **Flow** |
| Seed | 7 | Fund |
| Straw ploughed in | 0 | Fund |
| Straw burned | 7 | **Flow** |
| **Animal products and their utilization** | 95 | |
| Export from agriculture | 81 | **Flow** |
| Returned in offal | 13 | Fund |
| Milk for young animals | 1 | Fund |

Source: Data from the Institute of Agricultural Economics,
Denmark (1991).

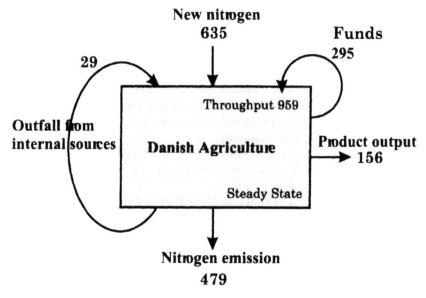

**FIGURE 7.7**
Nitrogen flows and funds in Danish agriculture in the middle of the 1980s. M kg N per annum. First level of analysis.

## Second Level of Analysis

To proceed to the second level, we can think of the agricultural production system in terms of three interacting subsystems: plants, animals, and fields (see Figure 7.8). To incorporate the unintended fund correctly, we have to add the system: Atmospheric outfall.

To establish how the subsystems interact, consider first the list of flows and funds in Table 7.4. Taking them from the top, the total fund of domestic fodder and bedding is the sum of the first two entries: 268+6 = 274 units. At this level of analysis, this becomes a flow from subsystem 1 into subsystem 2, see Figure 7.8. The fund "seed" (7 units) obviously remains as a fund, while the fund "straw burned" (also 7 units) becomes a (waste) flow.

Animal products returned in offal from slaughterhouses (13 units), and milk used to breed young animals (1 unit), remain as funds.

**TABLE 7.5**

The Nitrogen Balance of Danish Agriculture
(in million kg N per annum averaged
over a three-year period in the mid 1980s)

| | | | |
|---|---|---|---|
| FLOWS | | | |
| | A. | **Inputs** | **635** |
| | | Mineral fertilizer | 393 |
| | | Imported fodder | 170 |
| | | Biological fixation | 38 |
| | | Atmospheric outfall from external sources | 20 |
| | | Ammonia–nitrogen to straw | 10 |
| | | Municipal sludge | 4 |
| | B. | **Outputs** | **156** |

| | |
|---|---|
| Plant products | 75 |
| Animal products | 81 |

| | | | |
|---|---|---|---|
| A–B | **Nitrogen emission** | | **479** |
| FUNDS | | | |
| | C. | **Plant funds** | **281** |

| | |
|---|---|
| Domestic fodder and bedding | 268 |
| Returned plant products | 6 |
| Straw ploughed in | 0 |
| Seed | 7 |

| | | |
|---|---|---|
| D. | **Animal funds** | **14** |

| | |
|---|---|
| Returned animal products (offal) | 13 |
| Milk for young animals | 1 |

| | | |
|---|---|---|
| E. | **Product funds (C+D)** | **295** |
| F. | **Atmospheric outfall from internal sources** | **29** |
| A+E+F | **Throughput** | **959** |

The unintended fund (29 units) consists of two internal flows combined, one from the animal production system before landspreading (16 units), the other after landspreading (13 units). There are other agricultural sources of ammonia–nitrogen, but they are small in comparison.

The content of nitrogen in landspread manure has been estimated at 242 units, by the Institute of Agricultural Economics, Denmark (1991). This combines with a flow of waste fodder estimated at 63 units to yield a flow of 305 units.

The quantities thus fixed are underlined in Figure 7.7. The remaining seven unknowns are determined by the four balance equations and three additional pieces of information:

Of the total amount of nitrogen deposited from the atmosphere, a certain fraction (a4,1), about 50%, arrives at the root zone during the growing season being as effective as mineral nitrogen fertilizer.

Of the total amount of nitrogen taken up by crops, a certain fraction (a1,3) is contained in roots and stubble. This fraction is close to the value for non-fixing plants since the production of nitrogen-fixing plants in the 1980s was small in comparison. The fraction is estimated at 20%.

The soil pool consists of two subsystems: landspread manure (animal waste) and plant residues. The first component is the least effective in terms of passing on inorganic nitrogen to the root zone during the growing season. A small fraction ($b_1 = 0.2$ in the mid 1980s) of the nitrogen in landspread manure is presented to

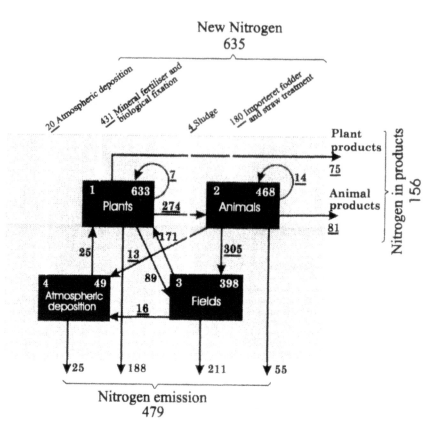

**FIGURE 7.8**

Nitrogen flows and funds in Danish agriculture in the middle of the 1980s. M kg N per annum. Second level.

growing crops with an effect equivalent to that of mineral fertilizer. A larger fraction ($b_2 = 0.8$) of the nitrogen contained in plant residues (including 63 units in waste fodder) is passed on to growing crops with an effect equivalent to that of mineral fertilizer.

This information translates into two equations:

*Equation 19    Fertilizer effect of ammonia outfall, Figure 7.8*

$$a4,1 = \frac{x4,1}{x4,1 + x4,w} = 0.5$$

*Equation 20    The root and stubble fraction, Figure 7.8*

$$a1,3 = \frac{x1,3}{x1,0 + x1,1 + x1,2 + x1,3} = 0.2$$

in addition to an estimate of $x3,1 = 0.2*242 + 0.8(4 + 89 + 63) = 173$ units.

These two equations and the estimate of $x3,1$ plus four balance equations, determine the seven remaining flows and all remaining COPs.

The software system displays the COP-equations in the window to the left of the graph area. In Figure 7.8, for example, the nitrogen efficiency in plant production (also called the fertilizer efficiency) defined as:

*Equation 21   The nitrogen efficiency defined, Figure 7.8*

$$a1,0 = \frac{x1,0 + x1,1 + x1,2}{x0,1 + x1,1 + x3,1 + x4,1}$$

is entered as it stands, and displayed on the screen by double clicking a1,0.

Having determined unknown flows and funds by means of the balance equations, we have at the same time determined the COPs at the time of the reference situation (the middle of the 1980s).

The analysis at the second level already begins to exhaust the available data. Nevertheless, the analysis should proceed to a level where the COPs become intelligible and operational. For instance, the efficiency in plant production defined by Equation 21 is not entirely operational, since it groups the input of mineral fertilizer with a small input of nitrogen from biological fixation. Clearly, these two inputs do not have the same effect. However, the error in doing so is small for the situation in the mid 1980s, since the biological fixation is only about 38 units out of 431 units entering subsystem 1 from the exterior. However, the model cannot serve as a good management tool unless we include nitrogen-fixing plants as a subsystem.

Likewise, the fodder efficiency defined as:

*Equation 22   The Fodder Efficiency, Figure 7.8*

$$a2,0 = \frac{x2,0 + x2,2}{x0,2 + x1,2 + x2,2}$$

is not operational since the efficiency is not measured as the nitrogen content of the products per unit of the total fodder input, but per unit of ingested fodder. There are, however, two additional waste emissions to account for, namely fodder conservation losses, and losses of nitrogen in animal waste.

### Third Level

To proceed to the third level of analysis, we detail the system in Figure 7.8 as shown in Figure 7.9.

Note that subsystem 1 in Figure 7.8 is the following subsystems in Figure 7.9: Cereal plants (non-nitrogen fixing), nitrogen-fixing plants, and straw. Likewise, subsystem 2 in Figure 7.8 becomes the following four subsystems in Figure 7.9: Fodder, animals, produced manure, and manure in storage.

Finally, it is clear that the third subsystem in Figure 7.8, fields, consists of two subsystems: landspread manure, and plant residues.

The flows, funds, and COPs are determined in a procedure similar to the one used in stepping from the first to the second level of analysis. First it is recognized that there are relations between flows and funds at the second and the third level of analysis. Outputs can be grouped into three categories: to the exterior as exported outputs, to other subsystems within the system, and to the exterior as waste. We need only to account for two of them since the balance equation determines the third.

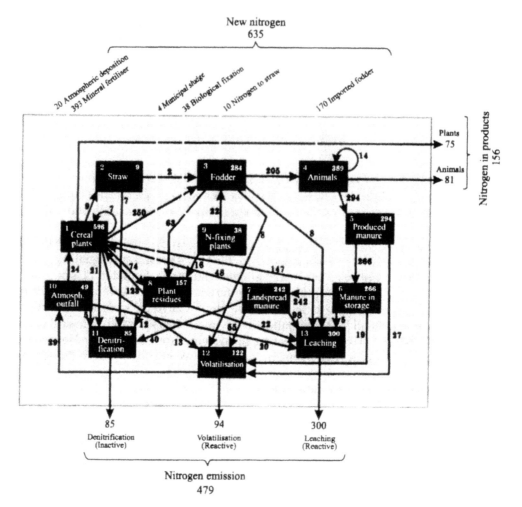

**FIGURE 7.9**
Nitrogen flows and funds in Danish agriculture in the middle of the 1980s in M kg Nper annum. Third level.

For example, the nitrogen flows across the boundaries of the subsystems Cereal plants, Straw, and N-fixing plants, combined (Figure 7.9) equals the fluxes over the boundaries of subsystem 1 in Figure 7.8.

Similar conditions apply to the three remaining groups and give a total of eight conditions. They ensure that the sum of the balance equations at the third level collapse into the balance equations at the second level when added over the relevant groups of subsystems. The result at the third level thus contains the result at the second level as a special case, just as the result at the second level contains the result at the first level.

The nine balance equations already available are thus supplemented to make 17 valid equations by which 17 out of a total of 30 unknowns are determined. As argued above, one can in addition add equations defining COPs and quantify them until the system is closed, and all flows, funds, and COPs are thus determined at the time of reference.

One can set up the equations and solve them to determine the unknowns, but it is faster to fix the flows determined by measurements and "tune" the COPs by trial and error until other measured flows are reproduced.

On the output side, only the distribution, not the total amount, depends on our ability to quantify the COPs. The total nitrogen waste output is fixed at 479 units at the first level of analysis and remains at that value at all subsequent levels of analysis.

At the third level of analysis, the COPs are operational. For instance, the two efficiencies of overriding importance: the efficiency in plant production (the fertilizer efficiency), and the efficiency in animal production (the fodder efficiency) are now defined as one normally defines them.

Here it is chosen to leave the matter of further detailing the nitrogen account, and complete the analysis by accounting for the form and origin of the emissions.

Nitrogen emissions enter the atmosphere as inert nitrogen and ammonia–nitrogen, or the hydrosphere as nitrate–nitrogen. To account for this, three logical systems (denitrification, volatilization, and leaching) are added as shown at the bottom of Figure 7.9. They merely group the nitrogen emissions into these three emission categories.

Data to allow estimations of the added COPs were provided partly by the Institute of Agricultural Economics, Denmark, and were partly estimated by the author based on previous work (Schrøder, 1985).

### The Danish Plan to Reduce the Nitrogen Emission from Agriculture by 50%

The plan to reduce the nitrogen emission from Danish agriculture by 50% is part of what is known as the *Action Plan for the Aquatic Environment* (Ministry of the Environment, Denmark, 1987) and amendments in 1991 and 1998.

The problem of rising levels of nitrate in groundwater and increasing frequency of occurrence of oxygen depletions in the seas around Denmark reached the agenda of Danish policy-makers in the 1980s. During a heated debate in Parliament in November 1986, it was decided to "prepare a plan, including budgets, to reduce the emission of nitrogen from Denmark by 50%, (and the emission of phosphorus by 80%) within three years."

This decision was somewhat modified soon after. For instance, the plan's time frame was extended from three to five years, and for some parts of the plan to seven years. In 1991, when it became evident that the plan was far from achieving the 50% goal, a slightly more rigorous plan was put into effect, extending the time frame to year 2000 (Ministry of Agriculture, Denmark, 1991). And in 1998 the plan again underwent political scrutiny and a new "second" plan was launched.

The phosphorus reduction goal has been achieved by application of the end-of-the-pipe approach, which applies for phosphorus since the major portion of that load comes from point sources. But this is the least important of the two nutrient loads. It has nothing to do with the problem of nitrate in groundwater, and very little to do with the recurring depletions of oxygen since nitrogen is the limiting nutrient in the sea around Denmark (Kronvang et al., 1993).

In countries like Denmark, agriculture dominates the national nitrogen balance. Danish agriculture has an input of about 635 M kg of new nitrogen per year, of which only about 30 M kg are found in domestic and industrial wastewater.

However, the plan allocated about ⅔ of the total budget of DKK 12,000 million (equivalent to about US$ 2,000 million) to reduce the output of nitrogen and phosphorus from point sources. The remaining budget (about 4,000 million DKK) was allocated to the task of reducing the nitrogen emission from the agricultural sector by 50%. A possible negative impact on the agricultural production was not considered.

The plan's policies, as far as the agricultural sector is concerned, can be summarized as follows:

1. Improved utilization of nitrogen in animal waste:
   - Facilities for storing liquid and solid animal waste are required to be able to hold the volume of nine months' worth of production.
   - Improved fertilization programs, including better timing of application of animal waste nitrogen.

- Improvement of animal waste application techniques.
- Animal waste is requested to be ploughed in within 12 hours after landspreading (tightened in 1991 to immediately after landspreading).
- Cover required on animal waste storage facilities to reduce volatilization.
- Elimination of "excess" fertilizer application (estimated at 10%).

2. Improved efficiency in plant production:
- "Green fields," that is, an increase of the area covered by crops sown during the fall.
- Discontinuation of the practice of burning straw in the fields. Excess straw is ploughed in.
- Improved fertilization techniques and change of crop composition to improve the nitrogen uptake of plants.

3. "Structural" measures:
- Reduction of the agricultural area.
- Promotion of organic farming.
- Fertilizer-free zones along rivers and streams.

4. International initiatives:
- Acknowledging that Denmark is not the only source of nutrient inputs to the sea, Danish policy-makers vowed to promote the idea of dealing with the eutrophication problem internationally. Today, countries bordering the North Sea and the Baltic have agreed, in principle, to reduce their nitrogen emissions by 50%.

### The Nitrogen Balance of Danish Agriculture 10 Years after the Plan's Implementation

The nitrogen balance in the middle of the 1990s is obtained by another calibration now using the result in Figure 7.9 as the starting point. The result is shown in Figure 7.10.

One can shuffle knowns and unknowns in the calibration process. In addition to the normal knowns (six inputs from the exterior) I have added the following based on information from Laursen (1993) and Danmarks Statistik (1992):

The effects of efforts under the heading "Better utilization of nitrogen in animal waste" are estimated to have had a twofold effect. First, the volatilization of $NH_3$-N from landspread animal waste has been reduced. Second, the utilization of nitrogen in landspread manure has increased from 0.2 to about 0.4.

The leaching of nitrogen from facilities storing fodder and manure has been reduced considerably.

The practice of burning straw in the fields is discontinued.

This adds six to the list of knowns. It is convenient to add a seventh variable, namely $x_{4,0}$, the net output of nitrogen in animal products.

According to Danmark's Statistik (1992), the output of animal products has increased by about 12% over the 10-year period after the implementation of the plan. Assuming that this applies also to the nitrogen output, then we have $x_{4,0} = 91$ units in the middle of the 1990s, as compared to 81 units in the middle of the 1980s.

Having thus transferred seven flows to the group of knowns one can tune relevant COPs until remaining flows fit independent observations. The result in Figure 7.10 is thus tuned

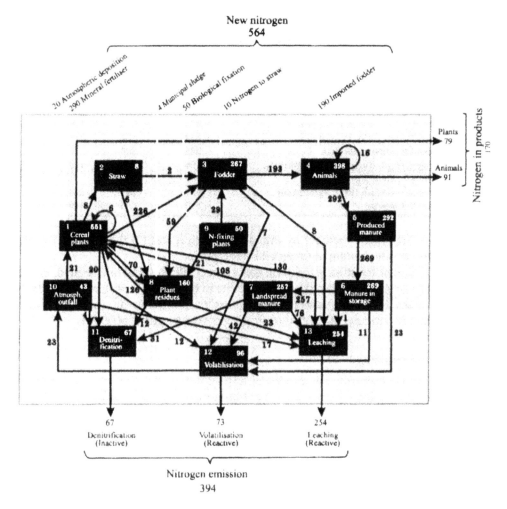

**FIGURE 7.10**
Nitrogen flows and funds in Danish agriculture in the middle of the 1990s in M kg N per annum 10 years after the plan's implementation.

to fit the following information concerning changes taking place from the mid 1980s to the middle of the 1990s:

1. The nitrogen taken up by crops and removed from the fields (x10+x11+x12+x13+x93+x90) has decreased from 363 units to about 348.
2. The total fodder intake has increased from about 389 units to 398 units.
3. There has been a slight increase in the net export of plant products.
4. The nitrogen contained in landspread manure has increased from 242 units to about 257 (6% as compared to a 12% increase in the output of animal products).

The increase of the fodder efficiency is confirmed by investigations reported by Sibbesen (1990) showing that the efficiency has increased by about 11% from 1980 to 1988, and by Damgaard Poulsen and Kristensen (1997).

The management options that have been changed from Figure 7.9 to Figure 7.10 are listed in Table 7.6. It is interesting to note that the two factors with the greatest effect on the

**TABLE 7.6**

Efficiencies and the Major Nitrogen Flows Around 1985, 1996, and at the 50% Reduction Goal

|                                        | Mid 1980s | Mid 1990s | Goal |
|----------------------------------------|-----------|-----------|------|
| Fertilizer efficiency                  | 0,57      | 0,58      | 0,64 |
| Fodder efficiency                      | 0,24      | 0,27      | 0,30 |
| Utilization of nitrogen in manure      | 0,20      | 0,42      | 0,70 |
| **New nitrogen (Mio. kg/annum)**       | **635**   | **564**   | **414** |
|   hereof in mineral fertilizer         | 393       | 290       | 115  |
| **Nitrogen in products (Mio. kg/annum)** | **156**   | **170**   | **170** |
|   hereof in animal products            | 81        | 91        | 91   |
| **Nitrogen emission (Mio. kg/annum)**  | **479**   | **394**   | **244** |
|   hereof leaching                      | 300       | 254       | 149  |

nitrogen balance are (1) the increase in animal production and (2) the increase in fodder efficiency, neither of which were anticipated by the plan.

The 15% reduction of the nitrate leaching has a direct, albeit small and lagged, effect on the nitrate content in groundwater. It has some effect in fjords (estuaries), but practically no effect in the open sea. In the international sea context, the ammonia–nitrogen volatilization must be included. It is the emission of chemically active nitrogen that counts.

### Accounting or Forecasting?

One may distinguish between two modes of application of the software, an accounting mode and a forecasting (or input–output modeling) mode. In the simplest case, the latter is a special application of the former.

In the accounting mode, the formulation of the COPs is unimportant. They may be formulated as standard output coefficients, or as physically meaningful COPs. They function merely as a convenient tool to set up energy-or-materials balances at given times of reference. The calibration exercises carried out to produce Figure 7.9 and Figure 7.10 are done in accounting mode although appropriate COPs are formulated.

In contrast, in forecasting, the "what-happens-if" mode, one must consider the validity of the two basic assumptions, namely that the system is in a steady, or quasi-steady state, and that it is linear.

In many cases, the steady-state assumption is acceptable. For example, energy systems react fast to changes in inputs and efficiencies, that is, the transient term of the conservation of energy equation is small compared to other terms. In such cases, quasi-steady modeling is acceptable.

Agricultural systems respond slowly to changes in inputs and efficiencies because of the long retention time of nitrogen in the soil subsystem, the humus pool. A steady-state model calculates as if the new steady state were attained instantaneously. The model jumps to the new steady-state solution that would be reached asymptotically after some time by a dynamic model.

The equations are linear. Plant growth experiments suggest that the nitrogen recovery function is curved at all application rates, but that the curvature from zero application to economic optimum is moderate. The Royal Society (1983) reports values in the range of 0.3 to 0.7 with an average close to 0.5 and note that "although recoveries decline as the level of fertiliser N is increased *above* the optimum application rate, they tend to be independent of the form of fertiliser applied, or the way in which it is applied."

In animal production, there is no good reason to believe that a nonlinear relation between the output of animal protein and the input of fodder protein should exist. The linear description is therefore reasonably accurate for forecasting purposes as long as changes are moderate. If changes are considerable, an iteration procedure can be used to correct for nonlinear effects.

### On the Feasibility of Reducing the Reactive Nitrogen Emission by 50%

So far the plan has had a marginal effect on the fertilizer efficiency, which can be said to be the bottleneck of the nitrogen account. However, the plan will not succeed unless this efficiency is increased considerably.

Suppose the goal were to reduce the emission of chemically active nitrogen to 50% of the level prevailing in the mid 1980s and, at the same time, to maintain the agricultural output of protein at the present level (1998). The following experiment suggests how much the efficiency in plant production shall be raised to meet this goal.

Based on the results reported by Sibbesen (1990) it is assumed that the efficiency in animal production can be increased about 11% above the level in the middle of the 1990s, and that the utilization of nitrogen in animal waste can be further increased from 0.40 to 0.70.

The options thus adjusted are shown in Table 7.6. The fertilizer efficiency must be increased from 0.58 to about 0.64 to achieve the goal. This is a substantial increase since the ideal (maximum) efficiency is about 0.77, at which the leaching of effective nitrogen is practically nil.

Having more or less exhausted the potential of other management options, the feasibility of reducing the active nitrogen emission from Danish agriculture by 50% (and at the same time maintain the present, 1998, agricultural production) therefore depends on whether or not it is feasible to increase the fertilizer efficiency.

### Is the System in Steady State?

If one applies a steady-state model to a known reference situation, it goes without saying that one implies that the system at the time of reference was in a steady, or nearly steady state. However, conditions in the mid 1980s may not have been in steady state.

In the 1950s, the input of fertilizer N was about 75 units, and the input of biological fixed N was about 260 units (Institute of Agricultural Economics, Denmark, 1991). A cereal crop leaves about 20% of its total nitrogen content in the root zone after harvest. A nitrogen-fixing crop leaves about 40% of the fixed nitrogen in the root zone (my estimate). Observing that the use of mineral N has grown fivefold from the beginning of the 1950s to the middle of the 1980s, and biologically fixed N has decreased to a mere 15% during the same period, it appears likely that the input of plant residual N to the soil has decreased. I estimate that it has decreased to roughly half of what it was in the beginning of the 1950s.

This substantial decrease has most likely resulted in a depletion of the content of plant residual nitrogen, and the depleted amount released from the pool of inorganic nitrogen in the root zone.

The rate of depletion of the stock of nitrogen could be in the order of 100 M kg per annum. Therefore, the total emission calculated assuming steady state for the mid 1980s is likely to be underestimated by approximately the same amount.

Up to about the beginning of the 1950s, agriculture had undergone slow changes over decades with gradual improvement in productivity. But the substantial decrease in the use of legumes, and the dramatic increase in the use of fertilizer N, taking place in a matter of three decades, most likely has changed the state of the system from a quasi- to a non-steady state. The content of nitrogen in the root zone, which has been increasing slowly but steadily over hundreds of years, has probably been decreasing since the middle of the century.

## Conclusion

The method enables its user to apply well-defined efficiencies and product allocation variables in input management of production systems.

Denmark's 10-year effort to reduce the nitrogen emission from its agricultural sector has so far produced marginal results. Leaching of nitrate–nitrogen has been reduced by about 15%. The chemically reactive emission (leaching and volatilization combined) has been reduced by about 17%.

It remains to be seen whether or not it is possible to increase the nitrogen fertilizer efficiency as much as required. A major improvement of fertilizer systems is required if the plan is to achieve its goal.

---

## References

Ayres, R. U., 1978, *Resources, Environment, and Economics: Application of the Materials/Energy Balance Principle*, John Wiley & Sons, New York.

Boulding, K. E., 1966, The economics of the coming spaceship Earth, in Jarrett, H., Ed., *Environmental Quality in a Growing Economy*, John Hopkins University Press, Baltimore, 3.

Carnot, S., 1824, *Reflections on the Motive Power of Fire and on Machines Fitted to Develop that Power*, Paris.

Daly, H. E., 1977, *Steady-State Economics*, Freeman, San Francisco.

Damgaard Poulsen, H. and Kristensen, V. K., 1997, *Normtal for husdyrgødning*, Danmarks Jordbrugs-Forskning, Beretning Nr. 736.

Danmarks Statistik, 1992, *Statistical Ten-year Review*.

Descartes, R., 1637, *Discourse on Method and the Meditations*, Penguin Books 1968.

Duchin, F. and Lange, G. M., 1994, *The Future of the Environment. Ecological Economics & Technological Change*, Oxford University Press, New York.

Georgescu-Roegen, N., 1971, *Energy and Economic Myths*, Pergamon Press, New York.

Georgescu-Roegen, N., 1971, *The Entropy Law and the Economic Process*, Harvard University Press, Cambridge, MA.

Hardin, G., 1968, The tragedy of the commons, *Science*, 162, 1243.

Hardin, G., 1985, *Filters Against Folly*, Penguin Books, New York.

Harremoës, P., 1995, *Water: A Precondition for Life*, Lectures from the 11th Convocation of the Council of Academies of Engineering Sciences, June 18-21 1995, Royal Swedish Academy of Engineering Sciences, Stockholm.

Harremoës, P., 1996, Dilemmas in ethics: towards a sustainable society, *Ambio*, 25, 6.

Institute of Agricultural Economics, 1991, Landbrugets økonomi, (in Danish with English abstract), Copenhagen.

Jørgensen, S. E., 1992, *Integration of Ecosystem Theories: A Pattern*, Kluwer Academic Publishers, Dordrecht.

Kronvang, B. et al., 1993, Nationwide monitoring of nutrients and their ecological effects: state of the danish aquatic environment, *Ambio*, 22, 4.

Laursen, B., 1993, (Institute of Agricultural Economics, Denmark), personal communication.

Leontief, W., 1966, *Input–output Economics*, Oxford University Press, New York.

Lovelock, J. E., 1988, *GAIA. A New Outlook at Life on Earth*, Oxford University Press, Oxford.

Miller, R. E. and Blair, P. D., 1985, *Input–Output Analysis: Foundations and Extensions*, Prentice-Hall, Inc., New Jersey.

Ministry of Agriculture, Denmark, 1991, *Action Plan for a Sustainable Development in Agriculture* (in Danish).

Ministry of the Environment, Denmark, 1987, *Action Plan for the Aquatic Environment* (in Danish).

Odum, E. P., 1989, Input management of production systems, *Science*, 243, 177.

Odum, E. P., 1971, *Fundamentals of Ecology*, Saunders College Publishing, Philadelphia.

Rifkin, J., 1981, *Entropy — A New World View*, Bantam Books, Toronto.

Schrøder, H., 1998, *Maximum Environmental Value for Your Money*, Danedi, Denmark.

Schrøder, H., 1997, *SteadyWin — A Software System for Environmental Management by Green Accounting*, Danedi, Denmark.

Schrøder, H., 1996, *Merging Economic and Environmental Input–Output Analysis*. Danedi, Denmark.

Schrøder, H., 1995, Input management of nitrogen in agriculture, *Ecological Economics*, 13, 125.

Schrøder, H., 1985, Nitrogen losses from Danish agriculture — Trends and consequences, *Agriculture, Ecosystems and Environment*, 14, 279-289.

Sibbesen, E., 1990, *Nitrogen, Phosphorus and Potassium in Fodder, Animal Production and Animal Waste in Danish Agriculture in the 1980s*, Report S 2054, Danish Ministry of Agriculture.

The Royal Society, 1983, *The Nitrogen Cycle of the United Kingdom*, London.

Ulanowicz, R. E., 1986, *Growth and Development. Ecosystems Phenomenology*, Springer Verlag, New York.

# 8

## Environmental Statistics

**Leif Albert Jørgensen**

## CONTENTS

## 8.1 Introduction

The aim of environmental statistics is to provide information on environmental issues on at least a nationwide basis by using methods normally applied to traditional national statistics. These include population statistics and National Accounting System, NAS, for calculation of gross domestic product, GDP. Environmental statistics involve a blend of the normal description of the society and the environmental issues of the society.

Environmental statistics can be divided to at least three topics.

A. A DPSIR model including the **D**riving force in the society, **P**ressure on the environment, **S**tatus of the different compartments of the environment, **I**mpact on society of the environment, and the **R**esponse from the society to environmental issues and problems.

B. A coupling between the National Accounting System, NAS, and calculations of emissions of specific compounds, a satellite accounting system.

C. An environmental use of statistical data gathered for other purposes, including the use of administrative registers. The data gathering can be changed, so that the normal data gathering by the Department of National Statistics or Bureau of Census can be utilized for collection of environmental information.

Some of the data-gathering systems of the Department of National Statistics will also be presented as well as the use of environmental information, since these systems often are unknown to the persons dealing with environmental issues and trained in natural sciences.

These are an overlay between the gathering of environmental statistics and the normal gathering of information about the environment by the Ministry of the Environment (Bundesamt für Statistik, 1997). The advantage of using the Department of National Statistics or Bureau of Census for gathering environmental information is the independence from the political system. The Ministry of the Environment has an elected person as its formal executive director, and statistics can too easily be manipulated to serve political directions of the elected top level. The Department of National Statistics has a long and traditional role of being independent from political issues formulated by changing governments. This is a crucial point for presenting independent information about the environment.

The presentation of environmental statistics as outlined can be very abstract if the presentation is done without illustrative examples. In the rest of this chapter the statistics will be presented with examples, covering the environmental problems of today. The presentation is based on the assumption that the reader should be a person trained in or with interest in natural and environmental sciences. This means that issues within social sciences and government administration will be described in a simplified way.

## 8.2   DPSIR

The DPSIR approach for organization of environmental statistics is widely used by the OECD, Eurostat, the national ministries of the environment and the European Environment Agency, for example (Stanners and Bourdeau, 1995). Also several of the publications from the Department of National Statistics (Statistics Norway, 1997) are organizing their environmental data according to this idea (UK Environment, 1992).

The basic idea of the DPSIR is illustrated in Figure 8.1. The system is a loop, but since the model is called the DPSIR the natural starting point will be the **Driving force**. The driving force is the factors in the society that are important for the environmental issue under consideration. The **Pressure** is normally the emission of pollutants from the society into the environment. The **State** is the concentration of chemical compounds in nature or the diversity of the ecosystems. The **Impact** is the way nature affects society causing deaths of humans or animals or changing the living conditions. The **Response** is the way society reacts to all of the changes in society or in nature to improve the conditions in the environment.

The DPSIR model is easier to understand when an example is applied as illustration. Figure 8.2 shows the DPSIR model used on fishery issues, creating environmental problems with overfishing and reduced fish stocks.

Figure 8.2 shows that the driving force on fisheries is the number of boats and the size of the fishing vessels combined into the fishing effort. The fact remains that if a fisherman is going to invest a surplus from his fishery, it will most probably be in a larger fishing boat. This creates a larger pressure on the fish stocks, where more hours can be spent on fishing and the number of fish caught will increase. The number of fish in the sea will be estimated by the Ministry of Fishery, and it is an important issue in describing the state of the aquatic environment. The impact of the changed environment is more difficult, but if a large number of cod are caught in the Barents Sea the seals in this ecosystem will look elsewhere for

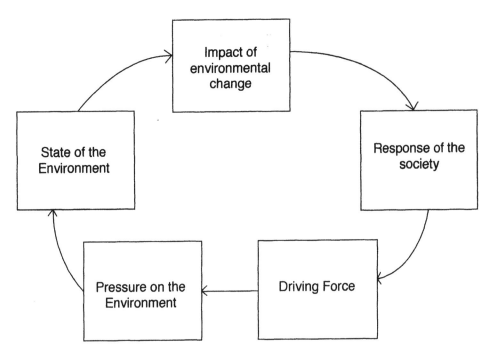

**FIGURE 8.1**
The DPSIR model for environmental statistics.

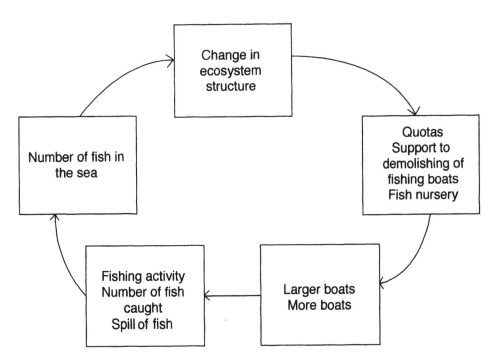

**FIGURE 8.2**
DPSIR model used for fishery.

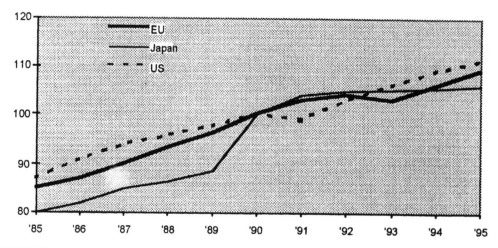

**FIGURE 8.3**
Growth of GDP per capita in different countries. Year 1990 = 100 units.

food, and the seals might migrate toward the coast of Norway. Society has many regulations for fishery, e.g., limiting the number of fish caught and giving support for reducing the number of fishing boats. Society is also giving support for the development of better fishing gear and electronic equipment and in this way counteracting the other regulations. The support from society for fishery and the tax paid by fishermen is easy to find in the National Accounting System and important numbers to describe the response of society to fishery.

Normal statistics dealing with fish and fishery issues are only considering the fish stocks and the landing of fish to give advice on future fish quotas (ICES, 1996). By using the DPSIR approach a more coherent system for providing statistics concerning the fish–man–society relationships can be presented.

### 8.2.1 Driving Force

Statistics for the driving force on environmental problems can be found in the normal statistics produced by the Departments of National Statistics. The question is only, what kind of numbers should be selected to describe a certain environmental problem?

The growth of the global economy is creating many of the environmental problems, both directly by depleting natural resources and by creating pollution during the different production processes. Closely connected to the depleting of resources is the growth of the population on Earth.

Figure 8.3 shows a strong growth in the GDP per person. The actual values are 17,243 EURO/person in EU, 25,374 EURO/person in the U.S., and 20,222 EURO/person in Japan. All values given for the year 1995. The growth will imply an increasing demand for most natural resources.

The driving force for many environmental problems is the demand for a larger output of the human systems and the structural changes in the production system to accomplish this change. It can be found in the agricultural sector, e.g., the larger farm units demanding increased use of capital, pesticides, fertilizers, and less use of man-power. It could also be the increasing demand for energy in society, creating environmental problems by building and using off-shore oil production facilities and coal-fired power plants. Also the deforestation in developing countries is a driving force for the destruction of the environment and can easily be described in quantitative terms as a driving force for many of the environmental problems related to biodiversity. It is a fact, that the natural ecosystems are not growing at the same rate as human society. This fact alone will create environmental and resource problems.

Another change societies are going through is the transition from an economy based on agriculture to an economy based on industrial production to an economy in which the service sector is the most important sector in the society. In Japan approximately 40% of the GDP comes from manufacturing, while in the EU countries and the U.S. only 32% and 25%, respectively, come from manufacturing. This change toward less of the GDP coming from manufacturing creates a society with less pollution, since much of the pollution is created by the manufacturing (Eurostat, 1997).

### 8.2.2 Pressure

In many contexts the pressure on the environment will be equal to the emissions. The emissions or deposits can be rather difficult to monitor, and the emissions may be taken as a measure of the pressure on the environment. Statistics or emissions to the air are well developed. However, there is a large difference between the emissions to the air and the subsequent deposition, especially when the emissions are from a large number of small emission sources or when the substances are transported long distances before deposition. So the pressure on the environment should not be the emissions, but the quantities that are added to the ecosystems on a specific time scale, e.g., the deposition rate of sulfur or mercury in an ecosystem.

Water ecosystems are more easy to deal with when the pressure on the environment must be determined. Modeling studies have for decades pinpointed this pressure as forcing functions to the water ecosystems. There can be difficulties when the pressure on a specific water body must be quantified. The emission from a sewage treatment plant is rather easy to measure, but this pressure will normally only be a part of the total pressure on an aquatic ecosystem. Let's consider a lake with a city and a sewage treatment plant. Some of the phosphorus will originate from the sewage treatment plant and some will originate from the streams flowing into the lake, and some will be deposited on the surface of the lake from wet and dry fallout. The last two are difficult to estimate. The problem gets even more complicated if an estuary or a part of the open Sea is considered. The phosphorus from the sewage treatment plants will be less important. There will also be a shift from phosphorus toward nitrogen as the limiting substance, when a more open marine water body is considered.

The way to overcome these ecological problems lies in the limiting substance principle is to consider the facts on a national basis and leave the more natural descriptions to the local administration. On a national level the total emissions from the treatment plants can be followed. This is shown in Figure 8.4, where the total emissions from all sewage treatment plants in Denmark are added together. The BOD is the biological oxygen demand.

Figure 8.4 shows a significant reduction of all measured substances from the Danish sewage treatment plants. The reduction has demanded significant investments in the treatment plants. The statistics presented in Figure 8.4 cannot tell anything about the pressure on the Danish aquatic environment, only the fact that the reduction of pollutants from sewage treatment plants has been significant. It also tells us something about the concern for the aquatic environment in Denmark. Pressure statistics can be more complicated when you are dealing with more complicated environmental problems. The problem with high ozone concentrations at ground level can only be described when a detailed knowledge about the formation of ozone is known. In northern Europe the concentrations of ozone are higher in

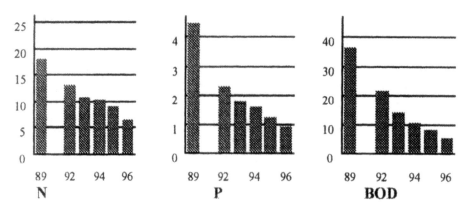

**FIGURE 8.4**
Emission of nitrogen, phosphorus, and BOD in 1000 of tons.

the countryside, because ozone in the cities is degraded by the air pollution. In southern Europe the highest ozone concentrations are in the cities. However, in both cases the emissions of ozone are zero — only reactions in the air will create ozone. An emission indicator could be the consumption or the emission of non-methane volatile organic carbon (NMVOC) and another indicator of the pressure could be the emission from the traffic in the cities. However, a single indicator for the pressure on the air quality considering the ozone problem is difficult to establish.

### 8.2.3   State

For persons with a background in natural sciences, environmental statistics will deal with the state of the environment by measuring concentrations. So, you just focus on a lake and measure the concentrations in the lake, e.g., phosphorus and nitrogen in the winter or the Secchi depth and phytoplankton in the summer. You should sample some polluted lakes and some clean lakes, and it would be preferable to have a geographical coverage, so a few lakes from each county would also be desirable. Some lakes have been measured for many years, so you could also include these in a nationwide monitoring program. A calculation of the mean should give the state of the environment in all lakes in the country.

However, this is the wrong way to select lakes to be used in a statistical system for monitoring nationwide water quality. Lakes should be selected randomly. If they are not, you cannot calculate nationwide changes in the state of the environment. If the monitoring program must cover these other demands about clean and polluted lakes, large and small lakes, lakes in different counties, you must include these factors when you are selecting your lakes. When an overview of the lakes and the selections are made before your sampling begins, you can calculate the nationwide state of the environment (Barnett, 1991). These conditions are too often overlooked when the selection of environmental locations are made.

### 8.2.4   Impact Statistics

The impact of pollution on the environment can be difficult to establish on a yearly basis. However, there are several cases where the damage to the environment is obvious. A good example is the damage in Indonesia and Malaysia from fires in the tropical forest in the fall of 1997. In Europe and the northwestern United States, acidification has also had a great impact on the economy, where acid rain has damaged buildings. These damages are economically important because restoration is expensive. In Denmark the damage has been estimated at some 100 millions of dollars per year. Impact of air pollution can also be found in reduced agricultural production, since the crops will be damaged by the ozone concentrations at ground level. In southern Europe the ozone concentration in the cities is so high that human health effects are anticipated.

This shows that there is a large demand for statistics on the impact of a changed environment. These statistics are not produced annually, because the impact is difficult to quantify on a yearly basis. The normal basis for the estimation would be a longer time. One country where environmental impact statistics can be produced is Tanzania, where the number of deaths from environmentally related diseases can be provided (the number of deaths caused by malaria and diseases related to bad drinking water).

Another impact is caused by the change in global temperature. Several independent institutions have tried to estimate the damage caused by global warming. There will be significant changes in the vegetation patterns, e.g., Arctic species will disappear from the Alpine region in central Europe. However, it is difficult to publish yearly environmental statistics in this field, because the investigations are based on longer time periods, and the change will take time and not happen from one year to the next. A measure of global warming could be based on statistics of the extreme weather conditions that have increased in frequency over the last decades. This might give an idea of the impact on the global scale.

### 8.2.5 Response Statistics

Response statistics show action from society on one or several environmental problems. The action can be of several different kinds, and the statistics produced must show this. The response statistics can be divided into the money spent, the number of convictions based on environmental laws, the number of new environmental laws, and the fulfillment of action plans.

The most obvious number concerning response statistics is the amount of money spent on environmental protection. The amount of money spent can be divided into governmental and local administration, and money spent by industry and private households. The amount spent by the administration can be easily obtained both from the national budget (amount planned to be spent) and from balanced accounts after the money has been spent.

Efforts in industry for pollution prevention could be assessed in the 1970s, when the frequent solution to an emission problem was the installation of a filter on the construction of a sewage treatment plant. Today the investment in environmental pollution prevention is almost impossible to measure. The reason is obvious. Engineers have been taught to integrate environmental protection into the technical design of the production equipment. These lessons have changed production equipment. New production units will not only have better economy, but will also be more environmentally friendly. A study of all green accounting for firms in Denmark in 1996 showed that not a single firm has invested only in environmental protection equipment. All investments have been integrated, so that a better production economy often may be more environmentally friendly. This is also the basic idea behind cleaner technology. The result is that it is not possible to calculate the investment in environmental protection equipment in industry.

Efforts have been made in the U.K. to compute the total environmental expenditure by industry (Brown, 1998). A postal survey of 10,000 companies was carried out. The questions covered end-of-pipe equipment, integrated clean processes, green products, in-house expenses, and current payments made to others. Other questions covered operational benefits and environmental employment. The result was that 0.5% of the GDP was for environmental expenditure by industry. Seven industries accounted for almost three quarters of environmental expenditure. This indicates that the environmental investments are happening with several years' interval and that the variation from year to year will be very significant. It could be much easier to cover only those industries that have large environmental expenditures. This would, however, demand prior information.

The expenses by households on environmental equipment is also difficult to summarize. The easy part is the money spent on waste collection and sewage treatment, since this amount is included in the municipal budget. This sum is a rather large amount, but it is not the only money spent on environmental protection. It is not possible on a national level to distinguish between clean and polluting products. For many years it has been desirable to account for, e.g., the sale of detergents with phosphorus and without phosphorus, but even today there are no indications of these sales. According to an investigation, 40% of the households are willing to pay 10% more for ecologically sound products from organic farming (Statistics Denmark, 1998b).

The criminal statistics are rather easy to collect through cases brought to court or treated by the police and carefully monitored by most societies. Old records can also be found. The numbers show that there has been rather little development in Denmark during the monitored period in the 1990s. The number of cases has been around 500 per year. The typical case is illegal disposal of waste or the problems in the agricultural sector of handling manure, burning straw, or applying pesticides. Even the average amount of money paid in each case has been constant, despite government demands for increasing fines for environmental crimes. The number of cases in Denmark is higher compared with neighboring countries. Both Sweden and Norway have a lower number of court cases. The number of cases brought to court is an important response indicator about the response of a society to environmental crimes. Legislation is not sufficient. There must also be a system to ensure the implementation and enforcement of the laws. Many countries have a low enforcement of environmental legislation. This can be found in several African states. The government is also active in many states in the private sector. This can create situations where the government will prosecute itself. This arrangement will make it difficult to enforce environmental laws.

The last type of society response statistics is statistics dealing with action plans. This type of planning is special not only from an environmental point of view but also in juridical terms. Action plans are often created by the Ministry of Environmental Protection or similar agency. During the formulation process several stakeholders will have a strong influence on the content of the plans. This will in the end lead to an action plan, where several stakeholders agree on the formulation of the plans. The content of the plans will, therefore, be vague on several points and from a purely statistical standpoint, the action plans will be difficult to monitor. Also the law creation process is strange. The Environmental Protection Agency or similar agency active in the environment will implement the plan, and it is the same agency that will monitor the progress of the plan.

An example of an action plan can be applied to illustrate the difficulties encountered. In Denmark, an action plan for reduction of the use of pesticides was launched at the end of the 1980s. Pesticides are considered important indicators of environmental response (UN, 1996). Several targets were firmly formulated; the use of pesticide should be lowered by 25% by the end of 1993, and a further reduction of 25% should be accomplished before the end of 1997. The reference period also was stated as the mean use of pesticides in the period 1981–1985 (Figure 8.5). This is all very clear. The plan clearly states that the weather and the number of pests should by included when the target is evaluated. How do you quantify the number of pests in terms of pesticides used? The results were that the number of pests and the weather conditions were of no concern. During the same period another action plan required the use of green fields during the winter period for reduction of the run-off of nitrogen. However, green fields in the winter will demand a higher dose of pesticides,

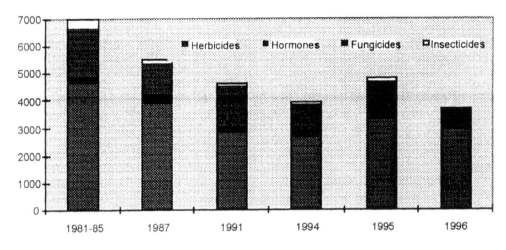

**FIGURE 8.5**
Tons of pesticides sold in Denmark.

because fungi will attack the green fields during the winter. This leaves the agriculture sector with two plans that have conflicting management development.

The use of pesticides was determined by demanding all importers, exporters, and producers of pesticides to give information about the tons of pesticides sold. However, the tons of pesticides sold will not match the tons of pesticides used by the farmers in the fields. This number should be obtained by contacting the farmers directly. At the same time taxes were imposed on the pesticides. This created a boom in the pesticides sold shortly before the taxes were imposed (1995) and a corresponding increase in the tons of pesticides sold. The next year the amount of pesticides sold was again lower.

Emergence of new pesticides and the ban of old pesticides are complicating the picture. The known harmful pesticides simazine and atrazine were among pesticides banned in the period, and new pesticides active in a lower dose (measured in g/acre) were introduced. This surely should have called for another measure in the action plan, not just the amount of pesticides sold. The number of times each field was sprayed with pesticides each year was introduced. Also this number should be reduced. This is a number calculated using a standard dose on each type of crop and of each type of pesticide.

The action plan was followed at the end of the 1980s by a tax on pesticides. The income of the government on this tax should be used for information and education of the farmers and for research in development of more acceptable pesticides or organic farming without use of pesticides. Also the effect of pesticides should be measured by the pesticide tax. When the pesticide tax was increased considerably later in the 1990s, this surplus was directed into the normal governmental budget. The money was not spent on initiatives that could reduce the use of pesticides. To summarize: action plans should be simple and have an agreed upon and measurable goal.

## 8.2.6 Covariables

Special attention must be taken when environmental statistics are used for natural systems or other systems, when the statistics are influenced by other variables at the same time.

A simple example is the concentration of mercury in fish. This concentration is often used to describe the state of pollution of a water body and is often detected in fish. However, the concentration in the fish increases with the length of the fish. If, by chance, all fish caught

tons N

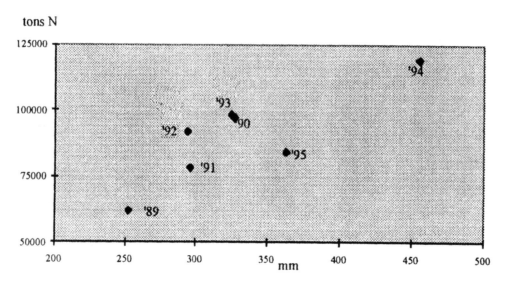

**FIGURE 8.6**
Nitrogen transported in rivers vs. runoff in mm per year.

are longer than in the preceding year, then the concentration of mercury would be interpreted as increasing, while the opposite might be true. The way to overcome this problem is to recalculate the concentration by a concentration vs. length relationship. A standard length of the fish could be selected to be, e.g., 30 cm.

Another example might be less obvious. The dissolved inorganic nitrogen concentration is the most important factor for the state of the marine environment and should be measured in a period when most of the nitrogen is in the dissolved inorganic phase, i.e., before the onset of primary production in the late winter in the temperate marine areas. This concentration could be used to describe the state of the marine environment. However, the transportation of nitrogen by rivers into the sea is controlled by factors outside the marine system. A very important factor in this transport is the precipitation and the volume of water transported in the rivers and streams. Normally one would expect that the concentration in a river would decrease with increased water flow. When the transport occurs from areas with a surplus of nitrogen in the soil, the opposite will happen. The concentration of nitrogen will increase with increasing water flow. The detailed mechanisms behind this relationship are not fully understood. Low concentrations will be found in meadow areas near the streams and higher concentrations in soil–water farther away from the streams. This gradient is observed in nature and can explain the rise in concentration when there is an increase in the water flow, because the water in the streams will originate from different soil–water pools. This indicates that the concentration of nitrogen in the marine environment reflects the pattern of precipitation in the period before the concentration is measured (Figure 8.6).

Time as a covariable in this last example. A seasonal pattern will almost always be found in natural systems. A standard way to compensate for seasonal patterns has been developed by the Statistical Bureau of Canada. The only problem in using this method for biological systems is that the seasonal adjustment demands equal time steps between the sampling times. This is impossible in most natural systems. Another mathematical method must then be used to interpolate between the sampling points to get a time series with equal time steps. This can be done by applying advanced mathematics, e.g., cubic splines or similar, but in many cases a linear spline together with a visual inspection of the interpolation offer the same quality.

**TABLE 8.1**

Economic Table for National Accounting, 1992

| No. | Sector | 1.1 | 1.2 | 1.3 | 1.4 | 1.5 | 1.6 | 1.7 | 1.8 | 1.9 | 1.10 |
|---|---|---|---|---|---|---|---|---|---|---|---|
| | | | | | | millions DKK | | | | | |
| 1.1 | Agriculture etc. | 3,452 | | 31,067 | 81 | 6 | 322 | 99 | 335 | 1 | 430 |
| 1.2 | Mining and quarrying | 59 | 35 | 27 | 4,147 | 10 | 187 | 3,044 | 204 | | 30 |
| 1.3 | Production of food, beverages etc. | 2,170 | 1 | 8,911 | 534 | 50 | 67 | 4 | 10 | 10 | 4,912 |
| 1.4 | Refineries, production of chemicals | 1,805 | 73 | 896 | 1,931 | 1,116 | 1,215 | 147 | 2,100 | 1,731 | 2,864 |
| 1.5 | Production of basic metals, metal products | 1,330 | 93 | 1,193 | 719 | 10,408 | 1,018 | 179 | 3,591 | 618 | 3,021 |
| 1.6 | Other manufacturing | 118 | 27 | 1,492 | 734 | 1,330 | 7,532 | 83 | 4,789 | 216 | 6,460 |
| 1.7 | Electricity, gas and water supply | 634 | 17 | 1,109 | 559 | 725 | 646 | 530 | 101 | 162 | 3,706 |
| 1.8 | Construction | 568 | 105 | 193 | 98 | 282 | 226 | 1,387 | 124 | 756 | 13,026 |
| 1.9 | Transport | 488 | 376 | 1,348 | 741 | 719 | 1,721 | 45 | 1,721 | 9,294 | 9,441 |
| 1.10 | Services, market and nonmarket | 4,804 | 237 | 4,870 | 2,524 | 7,581 | 6,112 | 868 | 9,383 | 4,223 | 45,756 |
| 2 | Households | | | | | | | | | | |
| 3 | Imports | 8,438 | 438 | 9,107 | 17,128 | 20,375 | 13,334 | 3,062 | 5,160 | 14,257 | 11,839 |
| 4.1 | Net taxes | -294 | 10 | -488 | 105 | 271 | 253 | 63 | 375 | -955 | 13,942 |
| 4.2 | Wages + profit | 22,695 | 16,531 | 20,488 | 9,730 | 24,655 | 16,795 | 7,765 | 20,456 | 33,177 | 235,123 |
| 5 | Total | 46,267 | 17,945 | 80,214 | 39,030 | 67,528 | 49,428 | 17,278 | 48,349 | 63,491 | 350,550 |

## 8.2.7 Satellite Accounting

The Departments of National Statistics are calculating the gross domestic product (GDP), the value of all products and services inside a country. There has for several years been a demand to correct this GDP for the destruction of the environment. The depletion of natural reserves and the destruction of soil and water should all be accounted for and subtracted from the normal GDP. The aim of the calculation will be to create a green-GDP. Major difficulties are built into such a calculation. Only a few will be mentioned here. A monetary evaluation of items that are not on the market will be necessary before a calculation of a polluted lake can be done. What is the value of a polluted lake, and what is the value of an unpolluted lake? It is even more difficult to estimate the value of the diversity in a forest. What is the value of a wild rhino? Evaluation methods involve willingness to pay. The cost of methods for prevention of pollution and also the cost for cleaning the polluted environment can be used as monetary measures (Bryant and Cook, 1992). These three methods are giving three different values and this makes an inclusion into the National Accounting System difficult.

The trend is now away from the green-GDP and more toward utilizing the National Accounting System for environmental purposes and building satellite accounts instead. Satellite accounts are accounting systems which present complementary information alongside the traditional accounts in order to highlight other issues, such as the environment.

The satellite accounting system has as its core the National Accounting System. In Denmark the total economy is divided into 130 sectors. For each sector the value of production is accounted for, together with the goods bought in the sector from other sectors. This gives an accounting system with 130 rows and 130 columns. Since it is difficult to present a table with 130 rows and 130 columns, a concentrated version is used. In Table 8.1 the Danish economy is condensed into 10 sectors, giving an accounting system with 10 rows and 10 columns.

Table 8.1 shows in row 5 that the total production in agriculture, etc., is 46,267 million DKK (7.57 DKK = 1 Euro, and 6.48 DKK = 1 US$). Agriculture is delivering 31,067 million DKK to production of food, beverages, etc., (that could be pigs) and is delivering 335 mil-

**TABLE 8.2**

Input of Energy in Different Sectors

| | Gas | Crude Oil | Fossil Fuel | Renewable Energy | Sector | Total Economy | | Air Emission |
|---|---|---|---|---|---|---|---|---|
| | | | PJ | | | Millions, DKK | | Tons |
| 1.1 | | | 44 | 2 | → Agriculture etc. | | → | |
| 1.2 | | | 4 | | → Mining and quarrying | T | → | T |
| 1.3 | | | 43 | | → Production of food, beverages etc. | A | → | A |
| 1.4 | | 332 | 59 | | → Refineries, production of chemicals | B | → | B |
| 1.5 | | | 19 | | → Production of basic metals, metal products | L | → | L |
| 1.6 | | | 35 | 4 | → Other manufacturing | E | → | E |
| 1.7 | 80 | | 288 | 32 | → Electricity, gas and water supply | | → | |
| 1.8 | | | 15 | | → Construction | 1 | → | 3 |
| 1.9 | | | 76 | | → Transport | | → | |
| 1.10 | | | 99 | | → Services, market and nonmarket | | → | |
| 2.a | | | 229 | | → Private consumption | | → | |
| 2.d | 58 | 186 | 218 | | → Exports | | → | |
| 2.e | 3 | 13 | 10 | 1 | → Other final demand | | → | |

lion DKK to construction (in row 1.1; that could be sand and stones). The internal deliveries in agriculture are 3,452 million DKK (that could be small pigs delivered from one farm to another). So all the rows give the deliveries from one sector to other sectors in society, including the same sector. At the same time the columns give the value of the goods the sector is buying. Agriculture is buying 634 million DKK from electricity, gas, and water, and 1,805 million DKK from refineries, production of chemicals (that could be chemical fertilizers). In row 3 the value of the import is given, and in rows 4.1 and 4.2 the net tax on the sector is given and the wages and profit in the sector.

This can be expanded with environmental information. The key information is introduction of energy balances. Each year an input of energy into different sectors of the society is accounted, and the energy consumption is determined both in physical amount and in economic terms by questionnaire. For the environmental accounting, the economic accounting is not the key point. The physical accounting is more important. All larger industries will receive a questionnaire, where information about energy form, energy storage, and the cost for the consumption of the energy must be filled out. It's a key point that this questionnaire is mandatory. The usage of the energy has also been divided into heating, transport, light and heavy processes. There are some 24 different types of nonrenewable energy covering the most important such as coal, crude oil, petrol, and electricity, but also more rare energy types are covered, such as brown coal, tar, etc.

In Table 8.2 only the four most important energy types are shown, including renewable energy. All the rows in Table 8.2 are used as input to Table 8.1 where the rows in Table 8.2 will be used as input to the same sectors as shown in Table 8.2.

The calculations of the emission of the different air emissions are very simple when the basic data-gathering process is done. Each energy type is multiplied by an emission factor for each sector. The actual calculation is done on a subsector level, dividing the 10 sectors and households into 130 different sectors. There are seven different compounds calculated for air emissions. These are $CO_2$, $SO_2$, $NO_x$, CO, $NH_3$, $CH_4$, and NMVOC. The technical emissions factors are found in the CORINAIR database and given by the Intergovernmental Panel on Climate Change (IPCC, 1997). The result is shown in Table 8.3. Each sector is shown, and the air emission from each sector can be found.

Some emissions are based on further calculations. The emission for cars and other vehicles are based on a refined system for vehicles including 60 different types of vehicles. This

**TABLE 8.3**

Air Emissions from Energy Usage

| Sector | | | $CO_2$ | $SO_2$ | $NO_x$ | CO |
|---|---|---|---|---|---|---|
| | | | 1000 tons | | | |
| 1.1 | Agriculture etc. | → | 2,934 | 9 | 36 | 20 |
| 1.2 | Mining and quarrying | → | 326 | 2 | 6 | 2 |
| 1.3 | Production of food, beverages, etc. | → | 2,433 | 9 | 6 | 4 |
| 1.4 | Refineries, production of chemicals | → | 663 | 10 | 3 | 2 |
| 1.5 | Production of basic metals, metal products | → | 681 | 1 | 2 | 6 |
| 1.6 | Other manufacturing | → | 2,640 | 11 | 6 | 7 |
| 1.7 | Electricity, gas, and water supply | → | 30,192 | 135 | 88 | 44 |
| 1.8 | Construction | → | 1,029 | 1 | 10 | 20 |
| 1.9 | Transport | → | 5,438 | 11 | 57 | 101 |
| 1.10 | Services, market and nonmarket | → | 3,364 | 3 | 20 | 69 |
| 2 | Households | → | 10,931 | 5 | 52 | 518 |
| 3 | Imports with air | → | — | 66 | 64 | — |
| | **Total load to air in Denmark** | | **60,632** | **370** | **553** | **793** |
| | Export with air | → | — | 172 | 266 | — |
| | Danish ships abroad | → | 9,265 | 169 | 256 | 22 |
| | $CO_2$ absorbed in plants | → | −5,258 | — | — | |

means that when you are using your car to go to the office, the emission from the car is accounted for in the household sector. When you are using the official car from the institute the emissions are counted in the "Services" sector. However, if you are riding on a bus or taking the subway to the office, the emission is accounted for in the "Transport" sector. This is not what you normally would consider as transport, but this is consistent with the National Accounting System.

Some of the emission factors have been changed over time, since the power plants have introduced techniques for cleaning the outlet air for $NO_x$ and $SO_2$. These plants will monitor the emission of $SO_2$ and $NO_x$ and can supply the system with actual data for emission in the different years. In the later years more data on plant-specific emissions have become available because of different green accounting systems that are made on all power plants with a capacity above 50 MW.

Some air emissions are not covered by merely looking at the energy consumption. At refineries the crude oil is cracked, and shorter oil compounds are created and emissions of different compounds are produced. This chemical process is included in the system, but it is not related to energy consumption. On the oil fields in the North Sea some of the gas is burned as a part of the production process. This burning will increase when the oil fields get older. The production of cement and other heating of $CaCO_3$ will also create an emission of $CO_2$ that is not only based on the consumption of energy. Emissions by cement production have been studied by IPCC and reevaluated several times.

Renewable energy consumption also gives rise to some air emissions. Burning of wood and straw is not contributing to global warming, because the $CO_2$ released just recently has been fixed by the photosynthetic process. The same is considered when burning waste material. However, there are also other emissions when wood and straw are incinerated. These are $NO_x$, $SO_2$, etc. These emissions are included in the system by the method already described, i.e., by multiplying the energy consumption by an emission factor.

The basic idea in the satellite accounting can be described as follows. The energy consumption in each sector is determined (Table 8.2), and is distributed into the total economy (Table 8.1), and emissions are calculated (Table 8.3).

The results show that the largest $CO_2$ emissions are from the sector consisting of electricity, gas, and water supply. This is not a surprise, since the generation of electricity and heat demands large amounts of fossil fuel. Denmark doesn't use hydropower for producing electricity, and there are no nuclear power plants. The indirect generation of emissions when industries are using electricity or distributed heat systems are calculated as an emission from the sector that is generating the electricity or hot water. The reason is that the National Accounting System is based on the goods sold in the different sectors. The electricity sector is selling electricity and hot water. The industries are selling their goods, such as pharmaceutical products or machinery. The goods are accounted for in the selling sectors not in the sectors that are buying goods, such as electricity. If the other approach were taken, i.e., accounting in the sectors that are buying the goods, all emissions were in fact created by the households. This leads to a rather contradictory emission calculation. If an establishment is changing from using a power generating system based on coal and instead is buying power from a large coal-fired power generating plant, the emission is moved from the manufacturing sector to the sector for generating electricity.

The second largest emission of $CO_2$ is created by households. Although district heating is widespread in Denmark, the use of oil and natural gas for heating gives a large emission of $CO_2$ (in addition to other air emissions). The emission of different compounds can be added to total emissions, and can contribute to global warming.

Another possibility in the National Accounting System is to follow the goods, and for each type of goods in the system an emission can be calculated. This will clearly state what kind of goods are creating environmental emissions. The big advantage of using the satellite accounting system is that all emissions throughout the life of a specific good will be summarized. If, for example, the emission of one ton of meat is calculated, this emission will include the emission from the slaughter-houses, the farm, and the transportation involved before the meat is consumed.

The National Accounting System also includes Danish ships sailing on the oceans. Denmark is a relatively large shipping nation. The income generated by shipping companies is included in the Danish National Accounting System and included in the Danish trade balance. This also creates a demand for the emission from the energy consumption from the ships to be included into the Danish satellite accounting system. This can be rather strange since the ships might be sailing between the west coast of the U.S. and Japan on a regular schedule. This calculation is shown in Table 8.3. There is a special emission factor for international sea transport. Some of the emission factors are larger, reflecting the fact that normal sea laws are more liberal, than the strict regulation on the mainland. In total 9% of the Danish $CO_2$ emission originates from the emission from this kind of shipping activity (Statistics Denmark, 1998a).

When all the calculations are carried out and the emission is integrated with the economy, as already described, a whole new world of possible analyses is opened. First of all, the economic prognosis tools that most countries have developed can use the data for emissions, when the economic prognoses are made. These prognoses have a major political importance, and it is now possible together with the economic prognoses to predict emissions. The emissions can be calculated from the overall emission factors or assumptions of a certain reduction of the emissions due to technical innovations or due to investment in either cleaner technology or in emission reduction equipment. A normal emission calculation made by the Ministry of Environment cannot be used with the economic modeling because most economic models are based on the numbers provided by the Department of Statistics or by the National Accounting System.

The importance of the different sectors in society concerning air emissions can also easily be projected. It is possible to pinpoint the sectors that contribute most to air emissions. This is now just a trivial task. It is also possible to allocate the importance of the export to the different type of air emissions. It is possible to calculate the air emission of different sectors by considering the number of employees in the sector, or the surplus in the sector, the total turnover in the sector, or the net tax that the sector is paying to the society.

One major problem in using this approach to the calculation of the emission is that a very large part of the emission is going into the sector electricity, gas, and water supply. This is from a more environmental point not true because this sector is only producing goods that will be used in manufacturing sectors or for final consumption in, for example, households.

## 8.3   Additional Environmental Statistics

In addition to the previously mentioned environmental statistics, the national statistical departments are gathering a wealth of information, that can be used for production of environmental statistics. The natural sciences are rather low in activity in the field compared with other disciplines. Large integrated databases have been compiled for fertility, occupational health, unemployment, etc. The basic idea is that the existing data will be utilized better, when specialized persons are examining the data from an environmental basis. However, it should be made clear that the statistics offices will never allow a publication in which persons or single establishments can be identified. Researchers will always have a demand for the most detailed information for the formulation and examination of hypotheses, but the demand from the statistical offices will be to keep the data anonymous, never allowing single persons or establishments to be identified. The struggle between the researchers, administration, and police has been intense on this point, but the national statistical offices have not yet given any information on a personal level out to the public. Analysis can be done on the individual level, but the results will have to be on a more aggregated level. In the following, some of the possible analyses will be shown.

Every person has a personal number. This number is given when the person is born or immigrates. This number is used all over in society. Payment of salary, social support, bank accounts, credit card shopping, and military service are just a few instances in which you will need a personal number. From a statistical point of view, the personal number system is a time save. It is obvious that the system can be misused. There is in Denmark an independent state control institute that gives permits to firms and governmental or local administrations to run a register encompassing personal numbers. It will normally be difficult to get a permit to erect a computer data system including personal number. If a golf club or a tennis club, for instance, wants a computer system with personal numbers, it will not get a permit, because it can easily use the names and addresses of the members without needing personal numbers.

Another large register in many countries is the Building and Construction Register. In this register information is kept about all houses and flats in the country. The addresses are often coded in the same way as the addresses in the personal number system. The buildings are divided into farm houses, flats, institutions, etc. There is also information about the technical installations such as piped water, toilets, electricity and sanitary installations. The register was created in the 1960s, because the government wanted to improve the housing standard in Denmark. Money was given to landlords and private home owners to improve technical installations. After the first oil crisis in 1973 money was also available to improve the insulation of houses and flats. To keep track of the money the register was also used.

From an environmental basis this database contains important information about the sewage connections in all Danish houses. There are codes showing if the house is connected to:

A. A public sewage system

B. A private sewage system

C. A sewage container in the soil

D. A sewage container for toilet wastewater and mechanical cleaning for the rest of the household water

E. A mechanical cleaning system with soil infiltration

F. A mechanical cleaning system with emission to running waters or the sea

G. A biological cleaning system with emission to running waters or the sea

H. Direct emission to running waters or the sea without any cleaning

I. Other type of installation

J. No installed water system

Based on the same address coding system in the two different registers, the personal number system and the Building and Construction Register, it is possible to allocate persons in all the houses and flats in Denmark. In other words, all persons in Denmark are placed in a house or flat, and all houses and flats are inhabited by one or several persons.

It is now possible to calculate the emissions from the rural population. The emissions from all larger sewage cleaning plants are already monitored by the Ministry of the Environment and Energy in a computer system for sewage treatment. However, there is no centralized information about the sewage in rural areas. More and more attention has been directed toward the pollution from the rural population. Large investments are planned, and these will be made by the private home owners.

Each person has an emission equal to one person equivalent, i.e., 21.6 kg BOD, 4.4 kg nitrogen, and 1 kg phosphorus each year. The sewage installations are given a treatment factor. The factor depends on the compound, i.e., there are different treatment factors for biological oxygen demand, nitrogen, and phosphorus. The main results are shown in Table 8.4.

The estimation of the standard derivation and similar estimates for uncertainties is theoretically very difficult to make on these populations. The reason is that the statistics are not made on a subsample of a population, but on the entire population in a country, and similarly on all houses and flats. This means that there is practically no uncertainty, and the true mean is known. At present no theoretical methods have been described to calculate the uncertainties from a total population count. Some estimates of uncertainties can, however, be made based on the problems encountered during the process, but these uncertainties are not related to the theoretical standard deviation.

In Table 8.4, 39,729 persons are without a permanent address. This expresses an uncertainty. The explanation is that the persons could be moving between addresses or they could be working abroad for a time. Another uncertainty is the fact that 97,000 houses and flats have no inhabitants. Of course, some houses will be empty, but the number is too large. The reason is that several houses are under construction or they are only used during holidays. Embassies create another problem, since they are covered by a house number, but the persons living there are not registered by the personal number system. This type of explanation in total covers 0.75% of the Danish population and 3% of the Danish houses and flats. This is the best estimate of the uncertainties by this type of sampling method.

**TABLE 8.4**

Number of Persons with Different Type of Sewage Installations and Emission from Rural Populations, 1st of January, 1998

| Type of Sewage Installation | No. of persons | Emission of | | |
|---|---|---|---|---|
| | | BOD | Nitrogen | Phosphorus |
| | | tons/year | | |
| Total | 5,275,121 | 10,107 | 2,340 | 534 |
| A public sewage system | 4,596,613 | — | — | — |
| A private sewage system | 34,161 | — | — | — |
| A sewage container in the soil | 17,647 | 339 | 69 | 16 |
| A toilet container and mechanical cleaning | 255 | 20 | 1 | 1 |
| A mechanical cleaning system with soil infiltration | 335,033 | 6,363 | 1,296 | 295 |
| A mechanical cleaning with emission to running waters or the sea | 166,543 | 2,753 | 721 | 164 |
| A biological cleaning with emission to running waters or the sea | 2,827 | 26 | 12 | 3 |
| Direct emission to running waters or the sea without any cleaning | 8,812 | 175 | 36 | 8 |
| Other type of installation | 69,424 | 431 | 205 | 47 |
| No information | 39,729 | — | — | — |
| No installed water system | 2,277 | 0 | 0 | 0 |

The emission from the nonrural population can be found in Figure 8.4

### 8.3.1 Emission of Phosphorus

Figure 8.7 shows that the emission of phosphorus is largely from the rural population outside Copenhagen (the white area on the right side of Figure 8.7). This is not a surprise, since more than 99% of the population in Copenhagen is connected to a sewage system. It also shows that a geographical distribution can be made from the information in the registers. A finer division can also be made. We could in principle get the amount of pollutants from each house, but this could not be published due to the strict regulation of the use of registers. The value is not available to the local administration.

In Table 8.4 special focus is placed on the pollution from the rural houses outside any sewage installation. In some municipalities the number of houses with "Other type of installation" represent a large percentage. This is typical of small rural municipalities. The local technical offices at the municipalities were contacted by telephone, and they explained what kind of installation the code covered. A cleaning factor could then be applied.

In conclusion, the previous method for estimating pollution by the rural population will be described. The local counties would from maps estimate the number of houses without any sewage system. The rural population was estimated to be 2.5 persons in each house. Some counties would use an estimate of 2.7 persons. No actual counts of the number of persons in each house was used. Since at least 10% of the population moves every year, the information on the actual number of persons is very important.

This is just one example of the usage of registers for environmental information. Using registers is cost effective and Statistics Denmark has produced a large number of statistics using a relatively small number of employees. Attention should also be given to the trade balance of different products. Environmentally dangerous products can to some extent be found in the trade balance numbers and possible damage can be foreseen. The numbers kept are always the economically important numbers. Difficulties will arise when statistics are produced for small quantities of compounds, e.g., mercury, since this compound is only used in quantities less than one ton. It is also difficult to get good estimations of the amounts of waste, because the economic importance of waste is rather low. Much information can be extracted about the agriculture sector, because it has good coverage in the statistical systems.

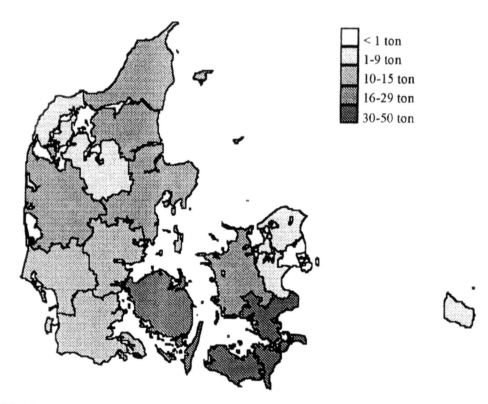

| | |
|---|---|
| ☐ | < 1 ton |
| ▨ | 1-9 ton |
| ▨ | 10-15 ton |
| ▨ | 16-29 ton |
| ■ | 30-50 ton |

**FIGURE 8.7**
Emission of phosphorus in the Danish counties to water from rural populations, 1998.

Another example is the use of materials in manufacturing. Each large firm receives a questionnaire about how much they are buying of different goods. Hazardous chemicals can be identified in the normal list of goods and a balance of the hazardous chemicals can be established. The advantage of using the department of statistics is also that when the reporting system is created, the data will be gathered annually.

### 8.3.2   National Nitrogen Balances

Nitrogen pollution is one of the major environmental problems in Denmark, and oxygen deficits are created in the inner Danish marine as a result of the nitrogen load (Hansen et al., 1990). A national balance for nitrogen can be created by combining environmental data and statistical data. The approach used is a mixture of economic/statistical models of the Danish agriculture sector. The data collection consists of the economic/statistical sampling of the yearly production of data from Danish farms. One third of all farms in Denmark get a questionnaire with at least 125 questions. It is mandatory to answer the questionnaire. The information collected is about crops, number of animals, machinery, irrigation, etc. (Table 8.5).

The input of nitrogen can easily be calculated. From the agriculture questionnaire and from national figures about fertilizers sold the total input of nitrogen to the agriculture can be calculated. The N-fixation by certain crops must also be added to the input. This is done by applying an area fixation rate for the specific crops that have nitrogen-fixation capacity and multiplying by the area. There is also a wet and dry deposition on the land surface. This value is determined by measurements by the Danish EPA (Kronvang et al., 1993). The nitrogen in the imported seeds is also calculated. The imported food for animals and the

**TABLE 8.5**

Input/Output of Nitrogen in the Danish Agriculture

|  | Total in Denmark | For Each ha of Farmland |
|---|---|---|
|  | t N·year⁻¹ | kg N·ha⁻¹·year⁻¹ |
| **Input** |  |  |
| Fertilizers | 326,000 | 121.0 |
| Biological N-fixation | 52,700 | 19.5 |
| Sludge | 9,700 | 3.6 |
| Deposition | 52,000 | 19.3 |
| Seed | 6,500 | 2.4 |
| Concentrates | 316,800 | 117.0 |
| Hay | 117,300 | 43.6 |
| Chemical foodstuff | 10,100 | 3.7 |
| Other | 13,100 | 4.9 |
| Total input | 904,600 | 336.0 |
| **Output** |  |  |
| Crop products | 341,500 | 126.9 |
| Animal products | 103,900 | 38.6 |
| Total output | 445,400 | 165.5 |
| **Loss** | **459,200** | **170.5** |

chemicals used for food for animals is also calculated on a nitrogen basis. Finally, the nitrogen content in the sludge from sewage treatment is added. Some of the wet and dry deposition will originate from the Danish agricultural emission.

The output of nitrogen from the agriculture products consists of the total production by crops and the total production by animals. The nitrogen content of milk, eggs, meat, etc., is added to give the total output of nitrogen from agriculture. The total loss from the agriculture sector can then be calculated as the difference between the total input and the total output. One major disadvantage of this method is the poor geographical coverage. The whole country is considered as one unit. A better geographical resolution would demand data on the regional use of chemical fertilizers.

A more specific model for the nitrogen loss from animals can be calculated from the same data source. A total of 39 coefficients are used and each coefficient is multiplied by the number of animals of a specific type. There is one type of horses, 19 types of cattle, 10 types of pigs, and 9 types of other animals. Some of the coefficients are changed from year to year. The loss can be geographically divided into the 14 different counties, and it is even possible to calculate the nitrogen loss from every specific farm or from a smaller defined area, such as a watershed. However, there are strict rules for privacy and confidentiality, and it is not allowed to publish data on a farm basis. The data must be summarized to give statistical information and not to provide information that could be used for administrative purposes. The loss of phosphorus and potassium can be calculated similarly.

The last model used is based on actual nitrogen measurements in rivers and streams in Denmark as a part of the Aquatic Environmental Nation-wide Monitoring Program. One of the aims of the national environmental monitoring system is to show the effect of the measures taken to abate pollution and to verify the fulfillment of action plans. Samples for nitrogen measurements are taken approximately every 14 days at more than 261 stations. The stations are chosen to cover the whole country and to cover different land use types, e.g., intensive farmland, forests, lakes, etc. The flow in the rivers is also measured, and the transport of nitrogen is calculated using the known relationships between concentration and river flow (Ministry of the Environment, 1997).

The loss of nitrogen from a land area can be calculated on an area basis from river transport and compared with the loss from agriculture and the loss from animals in the same

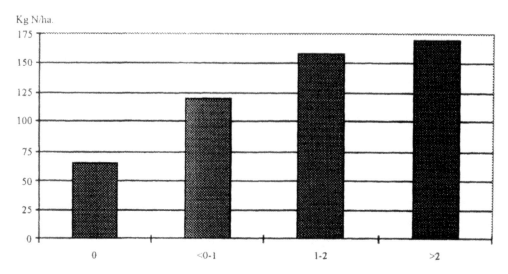

**FIGURE 8.8**
Measured loss of nitrogen in kg N/ha with different animal equivalents in the watersheds.

area. This is done by assuming that the nitrogen in the manure is utilized by a factor of 40%. The application of chemical fertilizers is adjusted for the nitrogen content in the manure on a county basis. The areas of the counties are then divided into nine watersheds.

The results show that the loss is approximately 171 kg N/ha farmland. The table shows also that on the input side the total input driven by the animals, i.e., the concentrates (foodstuff for animals, especially pigs), the hay and other foodstuff, is larger than the fertilizer input. The input of hay is from the import of hay and from fields outside the normal agriculture area. The output from crops is a factor of 3.2 larger that the nitrogen content in the animal products.

The loss of nitrogen from animals is important because the loss of nitrogen from farms with a larger number of animals is greater than the loss from farms without animals (see Figure 8.8). The loss is calculated per animal equivalent, and one animal equivalent is equal to one normal-sized cow.

The figure shows the importance of the number of animals in the runoff of nitrogen. It appears that the larger the number of animals in the watershed the larger will be the runoff. There is a large difference in not having any animals and in having a small number of animals. This shows that the problem with nitrogen loss is closely connected with the handling of manure and with the utilization of the nitrogen content of the manure.

It also shows the importance of the animals in the total turnover of nitrogen in the agriculture sector. There is a large difference in the geographical distribution of the loss of nitrogen from animals, reflecting the geographical distribution of animals.

Another measuring point in the nitrogen cycle is the runoff from the land area, measured in rivers and streams before entering the marine environment. The runoff calculation is divided into 9 principal divisions of the Danish land area determined by the watersheds and the receiving sea water bodies. In Table 8.6 is the nitrogen content in the runoff calculated and compared with the loss from farm animals and the total diffuse sources in the same areas. The total diffuse sources are calculated by adding the loss from the animals with the nitrogen fertilizers used. The geographical distribution of the chemical nitrogen fertilizers is done based on the farmland area in each watershed, assuming that 40% of the nitrogen content in the manure is utilized. This calculation will increase the use of chemical fertilizers in areas with few animals and decrease the use of chemical nitrogen fertilizers in areas with a high animal density.

**TABLE 8.6**

Nitrogen Runoff and Loss from Animals

| Area | Nitrogen with runoff | Loss from animals | Runoff/animal loss | Runoff/total diffuse sources |
|------|---------------------|-------------------|--------------------|------------------------------|
| | Tons N•year$^{-1}$ | | Pct. | |
| The North Sea | 21,000 | 91,400 | 23 | 10 |
| The Skagerrak | 2,250 | 3,080 | 73 | 32 |
| The Kattegat | 31,700 | 100,726 | 32 | 13 |
| Northern Belt Sea | 6,100 | 43,700 | 14 | 6 |
| The Little Belt | 7,900 | 28,000 | 28 | 12 |
| The Great Belt | 10,600 | 17,800 | 60 | 18 |
| The Sound | 1,850 | 4,100 | 45 | 9 |
| Southern Belt Sea | 750 | 3,600 | 21 | 5 |
| The East Baltic Sea | 2,250 | 7,200 | 31 | 9 |
| Total | 84,400 | | 36 | 13 |

Table 8.6 shows that only a minor part of the nitrogen in the manure and in the fertilizers applied to the land area are lost by the runoff to the sea. The fertilizer used in the watershed area is assumed to be equal to the national average corrected for the nitrogen content in the manure assuming 40% of the nitrogen content in the manure is used. The national average is that 13% of the total applied nitrogen is lost with runoff. Some of the variation in the importance of the point sources can be explained by the distance the freshwater has to travel before reaching the sea.

Calculations of nitrogen emissions from point sources, e.g., sewage treatment plants and industries with their own emissions show that the point sources are only 15.2% of the total nitrogen load in the runoff, including point sources to freshwater ecosystems. Even this value is an overestimate since some of the nitrogen in the freshwater point sources will be denitrified on its way to the sea. If only the point sources with emissions directly into the sea are considered, the load is 8.5% of the total nitrogen load in the runoff. The load from other countries (Sweden and Germany) and the deposition from the atmosphere directly on the sea surface should also be considered. This will reduce the importance of the point source further.

The nitrogen load from the point sources has been reduced 50% during the last 12 years. However, even before the reduction of the point sources, the nitrogen loads from these sources were only a minor part of the total nitrogen load from freshwater. The reduction at the municipal point sources has cost approximately 12·10$^9$ DKK or 100 DKK (15 U.S.$) for each kg of nitrogen removed in the sewage treatment plants.

The calculations give important values for nitrogen cycling through Danish society. The total loss from the input/output balance is 459,200 tons of nitrogen (171 kg N/ha.). The calculated loss of nitrogen from animals was 285,000 tons (104.7 kg N/ha.), and the runoff to streams was 84,000 tons of nitrogen in 1995, equal to 31 kg N/ha, if the same area equal to the agricultural area is used. If the total land area is used in the last calculation adding areas for forest and other areas outside the agriculture sector, the area loss coefficient for the runoff is 19 kg N/ha. The loss from the runoff does not include the evaporation of ammonia from manure.

The total loss of nitrogen from the input/output model gives the total strength of the diffuse nitrogen pollution sources in Denmark. This stress on the environment is approximately 170 kg N/ha. The geographical distribution of especially the animals must also be taken into consideration. This gives a maximum value of approximately 200 to 300 kg N/ha. This high value is in contradiction with the EU nitrate directive of 1999. The difference between the loss calculated from the input/output model and the loss of nitrogen from agriculture animals is caused by the fact, that the animal loss model only accounts for a smaller part of the nitrogen loss. The loss from the soil in the fields is not calculated, and the use of nitrogen in the manure is also not included in this calculation.

The trend in nitrogen transport toward the sea has not decreased in Denmark, in spite of all regulations, investments and change of agriculture practices. Some efforts have proven to be environmental contradictions, e.g., the regulations about having 65% green field in the winter season. These fields are not grass fields; the main part of the green fields in the winter have been winter wheat. This wheat demands a surplus of approximately 50 kg N/ha compared with a normal crop, thus allowing extra nitrogen to be leached out. At the same time, winter wheat demands a larger quantity of pesticides, since its more likely to be attacked by fungi. The advantage of the method used is that the input/output model and the animal loss model can be run with very little additional effort, once the model is developed.

The results show that only a minor part of the total nitrogen loss can be found in runoff. The open question is then, where has the rest of the nitrogen gone? Some of the nitrogen will be washed out and transported into the groundwater. However, it is impossible to find an increase in the concentration of nitrogen (nitrate) in the groundwater. Not because of lack of measurements, but because there are strict regulations about the quality of drinking water. When a groundwater well is polluted with nitrate or shows an increased level of nitrate, the drinking water works will change their mode of operation. Either the well will be closed or increased in depth until groundwater of better quality is found. In both cases, the physical conditions will be changed and the time series will be disrupted. Some of the nitrogen can also be denitrified in the soils, in the buffer zone around streams. The content of nitrogen in the soil can also change according to the use of the area. More focus should be directed toward processes for *removing* nitrogen from the farm soil before it ends up in streams and rivers.

## References

Barnett, V., 1991, *Sample Survey Principles and Methods*. Edward Arnold, New York.

Brown, A., 1998, U.K. *Environmental Accounts 1998*, ISBN 0-111-621022-2, pp. 75-82.

Bryant, G. and Cook, A., 1992, Environmental issues and the national accounts, *Economic Trends*, Nov. 1992, HMSO, London.

Bundesamt für Statistik, 1997, *The Environment in Switzerland. Facts, Figures, Perspectives*, ISBN 3-303-02034-5.

Eurostat, 1997, *Indicators of Sustainable Development*, Eurostat, European Communities, Luxembourg, ISBN 92-827-9827-5.

Hansen I. S., Ærtebjerg G., Jørgensen L. A., and Pedersen F. B., 1990, Analysis of the oxygen depletion in the Kattegat, the Belt Sea and the Western Baltic, Marine Research No. 1, Danish EPA, Copenhagen (in Danish).

ICES, 1996, *International Society for the Exploitation of the Sea*, Reports of the ICES Advisory Committee on Fishery Management 1996, Copenhagen.

IPCC, 1997, *Greenhouse Gas Inventory Reporting Instructions*, UNEP, WMO, WGI Met. Office, UK.

Kronvang B., Ærtebjerg G., Grant G., Kristensen P., Hovmand M., and Kirkegaard J., 1993, Nationwide monitoring of nutrients and their ecological effects, *Ambio*, 22(4), 176.

Ministry of the Environment and Energy, 1997, *Nature and Environment*, Copenhagen, ISBN 87-7772-375-9. http://www.dmu.dk/

Stanners, D. and Bourdeau, P., Eds., 1995, Europe's Environment. *The Dobris Assessment*, European Environment Agency.

Statistics Denmark and Eurostat, 1998a, *Danish NAMEA*, H. V. Jensen and O. G. Pedersen, Copenhagen.

Statistics Denmark, 1998b, *Environmental Statistics 1998*, Copenhagen, ISBN 87-501-1044-6.

Statistics Norway, 1997, *Natural Resources and the Environment*, Oslo, ISBN 82-537-4394-7.

*The U.K. Environment*, 1992, Department of the Environment and Government Statistical Service, London, ISBN 0-11-752420-4.

United Nations, 1996, *Indicators of Sustainable Development Framework and Methodologies*, New York, ISBN 92-1-104470-7.

# 9

## Life Cycle Assessment — Environmental Assessment of Products

**Michael Hauschild and Henrik Wenzel**

## CONTENTS

This chapter provides an introduction to the environmental assessment of products, its possible applications and the methodology used.

## 9.1 What Is Life Cycle Assessment?

Life cycle assessment is a methodology for assessing the environmental impacts and resource consumptions associated with the existence of products throughout the entire life cycle of the products — from cradle to grave. A central characteristic of life cycle assessment is thus the holistic focus on products (or the functions they fulfill) rather than on individual processes.

### 9.1.1   Introduction — The Life Cycle Perspective

The last decade of the 20th century has seen the emergence of a strong interest in the environmental impacts associated with the products that surround us and by which we obtain the many services that our civilization relies on. This interest has been accompanied by the development of methods for environmental assessment of products. Other environmental assessment schemes, such as environmental risk assessment of chemicals (ERA, see Chapter 10), or environmental impact assessment of major human activities, like construction works (EIA), focus on individual processes or groups of processes confined within one industrial installation. In contrast, the environmental assessment of products includes all the processes needed for the product to run through its life cycle — from the extraction of raw materials through production of the materials which are used in the manufacture of the product, to the use of the product and its disposal, with possible recycling of some of its constituents.

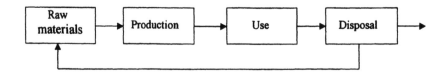

In contrast to most other environmental assessment tools, life cycle assessment is of a holistic nature in two ways:

- It attempts to cover the environmental impacts associated with the existence of the product regardless of *when* or *where* they occur
- In principle it attempts to cover all environmental impacts which are recognized as being significant today

As a consequence of the holistic product perspective, however, life cycle assessment is not capable of predicting actual environmental effects in the way that EIA and some forms of ERA do. This will be discussed later.

### *The Product System*

This agglomeration of processes, often separated from each other in time as well as in space, is referred to as the product system. Since the common feature for these processes is the fact that they interact with the product or its parts some time during the life cycle of the product, the environmental assessment of products is often referred to as *life cycle assessment* (LCA).

Life cycle assessment shares the focus on a process system with other environmental analytical tools like substance flow analysis (SFA) and material flow accounting (MFA) where focus is on the flow of a chosen substance (often a persistent pollutant such as one of the heavy metals) or a material within a given geographical region over a certain period of time. The object of these analytical techniques is the system of processes that the substance or material meets within the spatial and temporal boundaries of the study.

### 9.1.2 The History of Life Cycle Assessment

At the onset of environmental assessment of products back in the late 1960s and early 1970s, the life cycle assessment methodology was developed with strong inspiration from the already existing substance flow analysis. The first studies applying a life-cycle perspective to a process system took place in the U.S., focusing on environmental impacts from different types of beverage containers (Hunt et al., 1974). The name used for the assessment technique was *resource and environmental profile analysis* (REPA). Environmental awareness at that time was heavily influenced by the strong focus on resource depletion presented a few years earlier by the report: *Limits to Growth* (Meadows and Meadows, 1972) and by the experience of the first global oil crisis in 1973. Furthermore, the knowledge of the environmental consequences of anthropogenic activities was still too rudimentary at that time to allow a quantitative assessment of the impacts caused by the emissions from the product system. The focus in REPA was thus mainly on the consumption of energy and other resources.

The concept of environmental assessment of products led a quiet life throughout the 1970s. In the early 1980s much public attention in Europe was directed at the extensive use of resources for packaging of products. LCA experienced a renaissance through studies of the resource consumption and environmental emissions for different beverage container systems (such as beer cans and milk containers) performed in various European countries (Bundesamt für Umweltschutz, 1984; Lundholm and Sundström, 1985; Franke, 1984). In some of these cases it proved difficult to obtain reproducible results and conclusions, because the data and the methods applied varied between the different studies. This was unsatisfactory, and it spurred a more systematic development of the methodological basis for the environmental assessment of products.

From the end of the 1980s to today, interest in life cycle assessment has grown strongly, and an increasing number of different and often very complex products and systems have been assessed. Overviews and summaries of published LCA studies may be found in Pedersen and Christiansen (1992), *Sustainability* (1993), or Fullana and Puig, 1997.

### 9.1.3 The Interested Parties of LCA

According to Weidema (1997) the applications of LCA can be divided into four different types according to the time-perspective and intended use of the results, and the nature of the assessed products:

1. Noncomparative assessments of existing products
2. Comparative assessments of existing products (tactical applications)
3. Noncomparative assessments of potential products (product development)
4. Strategic assessment of products in relation to a strategic target

Society's groups of interested parties have seen different perspectives in the development of the environmental focus on products and have supported the work each from their own point of view.

#### Authorities' Interest in Life Cycle Assessment

Many industrialized countries adopted their first environmental legislation in the early 1970s, and for the first two decades the authorities focused most of their attention on the regulation of the problematic and hazardous chemicals and industrial installations (using tools of environmental risk assessment and risk management procedures, see Chapter 10). Through introduction of first cleaning technologies and later cleaner technologies in industry (see Chapter 6) this effort was in many cases successful. In Denmark, the total ecotoxicity impact from the wastewater discharged by the chemical industry was typically reduced by around 90% between 1980 and 1990 (Wenzel et al., 1990), and most municipal wastewater was biologically treated, thus relieving problems of oxygen depletion and ecotoxicity surrounding the discharge outlets.

While many local environmental problems were solved through regulation, the growing flow of materials and energy and the generation of waste continued to draw attention to problems such as global warming (mainly related to our use of fossil fuels), acidification (caused by combustion processes for energy generation), and photochemical ozone formation (to a large extent caused by fugitive emissions of volatile organic compounds from our transport and energy sector). The causes of these emissions and often also their sources are of a multiple and rather diffuse nature, which means that they cannot be controlled through regulation of the installations causing them. They are created due to our fulfillment of our needs, and a logical way for the authorities to address them is to focus the attention on the physical manifestations of this fulfillment — the products that we use. These products can be seen as responsible for their share of the emissions originating in the processes that enter into the product systems. In this perspective the products are the agents causing many of the diffuse contributions that add up to some of the most serious environmental problems facing today's society (Wenzel et al., 1997).

Several industrial countries including the European Union are discussing or have already implemented product-oriented environmental policies (National Environmental Policy Plan, 1989; Danish Environmental Protection Agency, 1996, 1998; Scholl, 1996; EC, 1999) aiming at reducing impacts from the products through a range of different measures.

- Ecolabeling or environmental declarations of products based on a life cycle assessment and thus reflecting the full environmental impact of the product. In Europe there are several national ecolabeling schemes and a common EU scheme covering a range of different product types
- Green public purchase guiding public purchasers in taking environmental considerations into account
- Take-back responsibility for certain product types (e.g., cars and electronics) making manufacturers liable to take their products back after ended use thus motivating them to design and construct the products with their disposal in mind

One future authority use of LCA may be to introduce a "green" taxation of products. An environmental tax which reflects the full environmental costs (including externalities) the product inflicts on society throughout its life cycle will cause the market to move toward the consumption of more environmentally friendly products (Danish Environmental Protection Agency, 1998).

Authorities can also use the holistic assessment principle of LCA in the environmental assessment of major societal action plans, of legislation, or more specifically of different ways of providing services like transportation, electricity generation, or waste treatment (Wenzel et al., 1997). For authorities, it is thus primarily application types 1 and 4 that are relevant.

### Industry's Interest in Life Cycle Assessment

To a certain extent, companies' interest in life cycle assessment tends to mirror the interests of the authorities. While the environmental regulation begins to include product-oriented environmental policies, some companies develop an environmental policy for their own range of products, possibly expanding from decisions in product development toward an integration of an environmental life cycle perspective into all decisions made by the company (also purchase, marketing, distribution, production).

So far, the core area for industry's use of environmental information from life cycle assessment is the product development (application type 3) where LCA is used for identification of the most important environmental aspects in the whole life cycle of the product, thus focusing improvement efforts where the problems are largest and setting targets for improvement (application type 1).

Another important use of LCA for many industries is in the marketing, recognizing that a certain (often growing) share of the market will prefer products that are appraised as "environmentally friendly" or "green" (application type 2). Environmental appraisal of products will often occur through application of an ecolabel or an environmental declaration.

### Consumers' Interest in Life Cycle Assessment

Consumers need relevant, reliable and intelligible information about the environmental properties of products in order to be able to take environmental considerations into account when purchasing. This information should be based on the environmental impacts caused by the product throughout its life cycle.

### 9.1.4   International Coordination of LCA Method Development and Standardization

Accompanying the growing activity within the field of life cycle assessment, much attention has been paid to the development of a sound methodological basis. The international scientific society of environmental chemists, SETAC (Society of Environmental Toxicology and Chemistry), started work on life cycle assessments in 1990 and has since then been the international forum for discussion of the methodological basis of LCA.

Although the discipline was still young and under development, the development of international standards for life cycle assessment was initiated in 1993 under the auspices of the International Standards Organisation (ISO). A general standard for the LCA area with ISO number 14040 was adopted in 1997, and more detailed standards for the different phases of the LCA are expected. These standards deal with:

- Goal definition, scope definition, and inventory analysis
- Life cycle impact assessment
- Interpretation.

## 9.2   How to Perform an LCA

In accordance with the present consensus within SETAC and in agreement with the current ISO 14040 standard, the life cycle assessment consists of the following phases:

- **Definition** of the goal of the assessment and scoping of the study
- Preparation of an **Inventory** of input and output (environmental exchanges) for the processes that occur in the product's life cycle

- **Impact assessment** where the results of the inventory are transformed into an environmental impact profile for the product system
- **Interpretation** of the impact profile according to the defined goal and scope of the study including sensitivity analysis of key elements of the assessment

Although described as consisting of four consecutive phases, life cycle assessment is an iterative procedure where experience gathered in a later phase may serve as feedback leading to modification of one or more earlier phases.

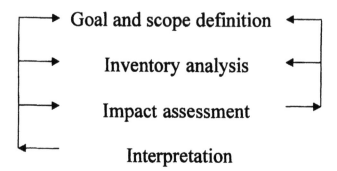

The four phases of the life cycle assessment are discussed in detail in the following sections. The review is primarily based on the methodology developed for the Danish Environmental Protection Agency under the EDIP program (Environmental Design of Industrial Products) and documented by Wenzel et al. (1997) and Hauschild and Wenzel (1998). The EDIP methodology is in accordance with the general recommendations in ISO 14040 and represents state-of-the-art method development within SETAC. In situations where there are important distinctions to other LCA methodology this will be stated explicitly in the text.

## 9.3   Goal Definition and Scoping

For the later interpretation of the LCA results, it is essential that the decisions that determine what kind of LCA is performed, defining the scope of the study, be explicitly stated in the assessment report. Any LCA report should start with an explicit declaration of the goal and scope of the study.

### 9.3.1   Goal Definition

The goal definition describes the purpose of the study and the decision process to which it shall provide input of environmental information. The definition of the goal serves in the later interpretation to qualify what types of questions the results of the LCA can be used for answering and, inherently, what types of questions it cannot answer.

The goal could be for a purchasing department in an organization to choose that product on the market which causes the least impact on the environment. Or the goal may be for the manufacturer of a specific product to identify the most important environmental impacts for all stages of the product's life cycle, in order to focus the effort to minimize the overall environmental burden from the product.

**LCAs are for <u>comparisons</u>, not for absolute statements**

In some situations it may be relevant to include the zero-alternative: What would happen if the product was not produced at all? This alternative (though often relevant) will rarely be included in an LCA conducted by a manufacturer of some new product. If the LCA-tool is used for evaluating public services like public transportation or large construction projects, it will be natural to include the zero-alternative.

For studies intended for publication or use in public decisions, the declaration of goals should also state the names of the different financing parties for the study. This will often play a decisive role in determining the scope of the study and may also influence the quality of data available, by having primary data available for some parts of the study.

As part of the goal definition it is important to state the possible extent of the decisions that will be made based on the study, as these will bear significantly on the definition of the scope of the study.


### 9.3.2   Scope Definition

Depending on the goal of the study, its scope must be defined in several aspects.


#### 9.3.2.1   *The Object of the Study — The Functional Unit*

In the first part of the scope definition the product is defined by the function it provides to the user. The object of an LCA is thus defined by the function or service it provides. This is due to the comparative nature of most applications of LCA. In order to ensure fair comparisons it is crucial that the product systems that are compared actually provide the same function to the user.

**LCAs are focused on functions**

Based on the function of the product, a *functional unit* is defined in quantitative terms, possibly complemented by a description of more qualitative aspects of the function. The functional unit defines the amount, contribution, or delivered effect (dependent upon the purpose and function of the product type) of the different products that shall be compared in the assessment. The choice of functional unit will often be very important for the outcome of the LCA comparison of alternative products, and therefore its definition should be considered carefully.

The definition of the functional unit is straightforward for products for immediate consumption, but will often require careful analysis for durable goods. For example, in a comparison of cartons and glass bottles for milk packaging the functional unit was defined as "packaging of 1000 liters of milk" (Lundholm and Sundström, 1985). For a refrigerator it might be "a volume of 200 liters refrigerated at 5°C for 13 years at an ambient temperature of 25°C" (Mose et al., 1997). In order to ensure fair comparisons, the quantitative description of the functional unit may have to be supplemented by a description of important secondary qualities of the product that are essential for the user's perception of the products as being of equivalent value. For the refrigerator the description of the secondary values could be: "It is self-defrosting, it has evaporation of water from defrosting, and it has a door and a number of shelves, baskets, and boxes as shown in drawing (accompanying the scope definition). These functions have a durability which corresponds to the life span." (Mose et al., 1997).

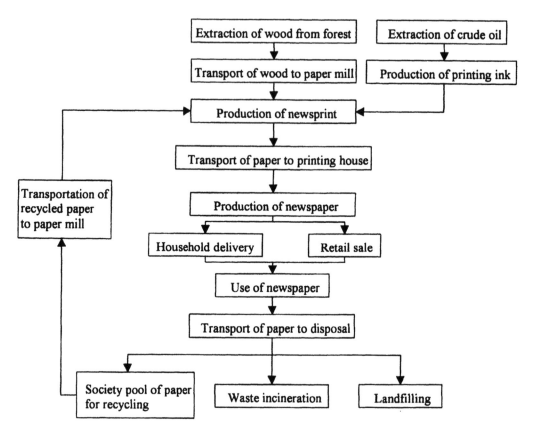

**FIGURE 9.1**
Process tree for the life cycle of newspapers.

### 9.3.2.2  *The Product System*

A central part of the scope definition is the elaboration of a thorough *definition and delimitation of the system studied in the assessment*, including:

- Detailed description of the life cycle of the product and drawing of boundaries between the product system and the environment
- Specification of the individual processes including the extent to which the product draws on them in its life cycle

For simple products involving few materials and processes the product system is often depicted by a process tree as shown for newspapers in Figure 9.1.

When defining the product system it will always be necessary to omit some processes that in principle do contribute to the product system in order to make the subsequent inventory analysis feasible. If a product system involves, e.g., a flow injection process, apart from the environmental exchanges from the flow injection processing of the product, a proportional part of the environmental impacts from the production of the flow injection equipment should be allocated to the product. If the flow injection equipment is made from steel, a proportional part of the impacts from the construction of the steel plant should be allocated to the production of the flow injection equipment and a share further on to the product, and so on, and so on *ad infinitum*.

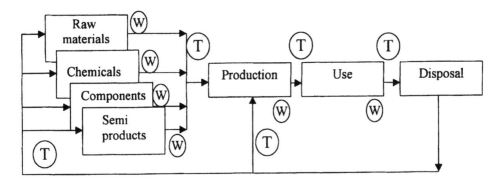

**FIGURE 9.2**
Life cycle of a product. Transportation processes are shown by a circumscribed "T", waste treatment processes are shown by a circumscribed "W".

Very quickly the product system expands to comprise the major part of human activities since the start of civilization, and generating an inventory for such a product system is an impossible task. Fortunately, it is also an irrelevant task. Most of the processes thus included in the product system will not contribute significantly to the impacts of the product. In general, the further away from the product, the smaller the contribution. Very often, it is thus irrelevant to include even the capital equipment in direct contact with the product proper.

A rule of thumb in defining the product system is that

> **only those activities that are relevant (i.e., can change the final result of the assessment to such an extent that they will influence the decisions that will be based on the outcome of the study) shall be included in the product system**

This is the paradox of the scoping of the product system. Only the relevant parts shall be included, but to determine which activities are relevant, you have to scope the product system and perform the life cycle assessment. To circumvent the paradox the scope definition of the product system is performed through an iterative procedure where it is influenced by feedback from the inventory analysis, the impact assessment, and sensitivity analysis phases of the LCA.

In the reporting of the LCA, the definition of the product system shall *explicitly state where the boundaries of the product system are drawn and substantiate this through sensitivity analyses.*

### Life Cycle Stages of the Product

Figure 9.2 gives a diagrammatic presentation of the stages in the life cycle of a product. Each process in the life cycle receives inputs of resources, materials, and energy and produces:

a. Outputs of materials, components, or products to subsequent processes
b. Emissions to the environment

Aspects associated with the scoping of the different life cycle stages are:

**Extraction of raw materials.** This stage in the life cycle includes the extraction of all materials involved in the entire life cycle of the product. Typical examples of activities included in this stage are mining of ores and minerals, forest logging, crop harvesting and fishing. If indicated by sensitivity analyses, the inventory for the extraction of raw materials must include raw materials for the production of the machinery (capital equipment) involved in the product life cycle. Raw materials used in the generation of electricity and energy used in the different life stages of the product should also be included.

It is still a common error in LCA to leave out relevant parts of the raw materials stage from the assessment. Often, some quite substantial environmental impacts of the product life cycle will be associated with this first stage. The decision of what to include in the LCA and what to leave out should thus be based on a sensitivity analysis.

**Manufacture of the product**. The production stage encompasses all the processes involved in the conversion of raw materials into the product considered in the LCA. Apart from the manufacturing processes at the plant where the product is made, it also includes production of ancillary materials, chemicals, and specific or general components at other plants, often located elsewhere in the world.

**Transportation**. As indicated in Figure 9.2, transportation is really not a single life stage in itself. Rather it is an integral part of all stages of the life cycle. Transportation could be characterized as conveyance of materials or energy between different operations at various locations. The transportation may also have to include an appropriate share of the environmental exchanges associated with the construction and maintenance of the transport system, whether this be road, rail, water, or air transportation.

**Use of product**. The use-stage of the product occurs when it is put in service and operated over its useful life. It begins after the distribution of the product and ends when the product is used up or discarded to the waste management system. Included in the use-stage are environmental exchanges created by the use or maintenance of the product.

**Disposal**. Waste may be generated by processes in any stage of the life cycle and, as such, waste treatment processes may occur in any stage of the life cycle (like transportation processes). Since management of the product's disposal is often one of the key aspects of the management of the product system, the final disposal of the product is normally given a stage of its own.

The treatment of the waste may involve alternative processes such as:

- **Reuse**. Use of the product or parts thereof in new units of the same product or in different products
- **Recycling**. Use of materials in the product itself or in other products
- **Incineration**. Combustion of the product, generating heat that may be used for electricity production or heating
- **Composting**. Microbial degradation of biological materials yielding compost for improvement of agricultural soils
- **Wastewater treatment**. Degrading organic matter and removing nutrients from sewage water, creating sludge that is deposited on agricultural land
- **Dumping**. Deposition of the product or its parts in landfills

Each of the forms of waste treatment mentioned may be considered as a processing of waste associated with a certain consumption of resources. This results in various emissions to the environment, and the possible generation of energy or of materials that will be an input to other processes in the life cycle of this product or other products.

### 9.3.2.3 Assessment Criteria

Dependent on the goal defined for the study, the criteria on which the product system(s) shall be assessed must be determined as part of the scope definition to ensure that the data collected for the product system during inventory analysis are relevant for the chosen assessment criteria.

Though there have been attempts to include socioeconomic and ethical aspects as assessment criteria, the majority of studies only consider direct environmental impacts and resource consumptions as assessment criteria. In addition, particularly the Scandinavian countries also include impacts in the working environment for operators of the processes. The EDIP method allows inclusion of a range of mainly chemical-related working environment impacts (Wenzel et al., 1997).

SETAC has provided a guiding list of environmental impact categories to include as assessment criteria when performing an LCA.

| Global Impacts | Regional Impacts | Local Impacts |
|---|---|---|
| Global warming | Photochemical oxidants | Odor |
| Ozone depletion | Acidification | Destruction of landscapes |
| | Nutrient enrichment | Division of habitats |
| | Human toxicity | Radiation |
| | Ecotoxicity | Accidents |

Source: Udo de Haes, H., Ed., 1996, *Towards a Methodology for Life Cycle Impact Assessment*, Society for Environmental Toxicology and Chemistry — Europe, Brussels. With permission.

Deviations from this list should be reported and explained.

### 9.3.2.4 Time Scale

Scope definition must also address the time scale of the study, particularly the requirements on the future validity of the results, i.e., for how long shall the conclusions of the study be valid. This may have a profound significance for the choice of technology for the processes in the product system and for the data that are collected during inventory analysis. If the LCA is intended to be used in product development for durable goods, such as a television set, the conclusions must be valid for several years, maybe even a decade. For the development of ecolabeling criteria which are based on a life cycle assessment, the recommendations must be valid as long as the criteria are intended to be in force, typically 2 to 4 years. An LCA used for environmental declaration of a product must in principle be valid for as long as the product is in use. For consumer goods such as food, this may be a very short period.

### 9.3.2.5 Technological Scope

Definition of the technological scope involves determination of the relevant technology for all the processes in the product system: average technology, specific technology, best available technology, predicted technology? The choice follows from the scope of the study. For example, an LCA may be performed to provide some background for decisions to be made during development of a new and more environmentally sound product. In this case, the technology assumed for the different processes should be that which is present at the factory, and intended to be used for the manufacture of the new product. If, on the other hand, the purpose of the LCA is to develop criteria for awarding an environmental label to products within a certain product group, then all existing levels of technology used to manufacture this type of product within the relevant region should be considered. The criteria would then be based on the functioning of the best technology available, so that only the least contaminating products would receive the label.

For long-lived products in particular, it may be relevant for the processes occurring late in the product life (use-stage or disposal stage) to attempt to predict trends in the development of the relevant technologies, such as energy generation or waste disposal technologies.

Figure 9.3 shows examples of trends in processes from the different life cycle stages of products.

### 9.3.2.6    Allocation Models

Some of the processes occurring in a product system will provide more than one marketable or useful output and hence enter into the product systems of other products as well. For processes that enter into more than one product system the environmental exchanges must be shared between the relevant products. This is the task of allocation, to identify joint processes in the product system and to determine a key for allocating the environmental exchanges of the joint processes between the products (Figure 9.4).

The problem of allocation between a product and by-products of a process is almost inevitable in the chemical industry, where many processes yield more than one useful product. Indeed, many new materials have been developed from substances that were originally considered waste products from other processes. Other examples are provided in agriculture where the breeding of cattle provides both meat, hides, and manure as output, or in electricity generation by combustion of fuels or through nuclear power where there is a concomitant production of (waste) heat.

A different allocation problem occurs in product systems that involve recycling loops, particularly the most common type of recycling, the recycling via open pool known for metals, paper, and glass in many countries. With this type of recycling the product is collected and its content of recyclable materials extracted into a pool that provides raw materials for products that may be unknown and generally belong to other product systems. An example is the use of paper-fibers from recycled office paper as a raw material for tissue paper or disposable diapers.

It has proven difficult to find generally accepted solutions to the allocation problems, partly because of the artificial nature of the boundaries that we, in our attempt to relate the environmental impacts to individual products, invent between our product system and the conglomerate of other product systems that surround us and jointly constitute our technosphere. In the real world environmental impacts are associated with the processes that cause them. Nonetheless, an internationally agreed procedure has emerged recommending to solve the problem by:

1. Avoiding the allocation by expanding the product system to include its additional functionalities. The function of cattle-breeding is thus not only to produce meat but also to produce manure. If cattle-breeding is compared to other ways of producing protein for human consumption, the alternative product systems should thus be expanded to include production of fertilizer with a fertilizing value equivalent to that of the manure and materials that can replace the hides. It is obvious that for more complex product systems the expansion is not a realistic option because it requires too much additional work collecting data about systems that are really without interest for the problem under study. Furthermore, the additional uncertainty thus introduced into the inventory may be significant.

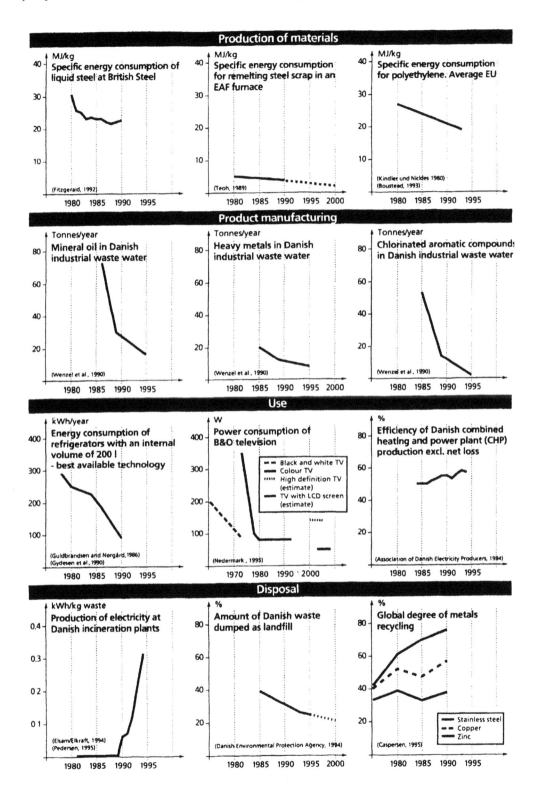

**FIGURE 9.3**
Prediction of developments in some key features of the different stages of products' lives (From Wenzel, H., Hauschild, M., and Alting, L., 1997, *Environmental Assessment of Products, Volume 1, Methodology, Tools and Case Studies in Product Development*, Chapman and Hall, London. With permission.)

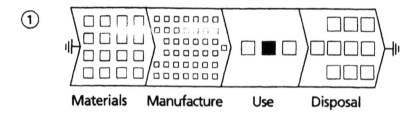

Materials    Manufacture    Use    Disposal

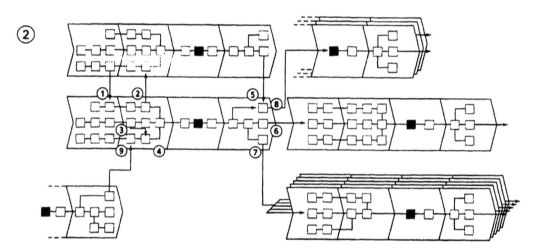

**FIGURE 9.4**

(1) Diagrammatic presentation of the product system. (2) Interrelations between the studied product system and other product systems. (From Wenzel, H., Hauschild, M., and Alting, L., 1997, *Environmental Assessment of Products, Volume 1, Methodology, Tools and Case Studies in Product Development*, Chapman and Hall, London. With permission.)

2. If system expansion is not possible, allocation between the different product systems should be performed using a technical criterion for the utility value that the process provides to each of the product systems. For the cogeneration of electricity and heat the common technical utility criterion may simply be the energy content in MJ of electricity and of heat.

3. In some cases it is not possible to identify a technical criterion that is relevant for both product systems. In this case it is recommended to base the allocation on an economic criterion, typically the relative price of the outputs. A classical example is the mining of diamonds, which has (at least) two outputs: raw diamonds and road material. It is not possible to identify a common technical utility criterion of the two products and therefore the environmental exchanges of the diamond mining process shall be allocated based on their relative prices (which incidentally also seems to be a fair allocation key).

## 9.4   Inventory Analysis

For each of the processes within the product system defined during goal and scope definition, information is now collected on the input and output (environmental exchanges) and possibly on the internal interactions with an operator if working environment is to be included in the impact assessment.

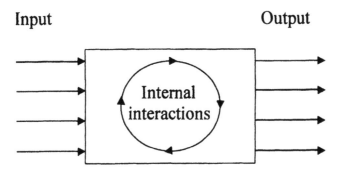

The compilation of inventory data *relates the inputs and outputs of the different processes to the life cycle of the product.* This quest for product specificity is a very important point. It is a fundamental characteristic of the life cycle inventory and the ensuing impact assessment, the purpose of which is to evaluate the environmental impacts of *the life cycle of the product.*

### 9.4.1 Collection of Data

In general, the collection of data is based on mass balances for the process over a longer period of time. It is important to ensure in this way that the data are representative of the average functioning of the process and that irregularities in the service, such as start-up and closure, cleaning of equipment, etc., are included.

The environmental exchanges must be accompanied by information on their statistical uncertainty. The data are reported as the process' environmental exchanges per functional unit. In the reporting of the inventory, the exchanges are generally aggregated and presented for the different life cycle stages as well as for the entire life cycle.

Traditionally, the emissions are divided according to the environmental "compartment" into which they are released:

- Air (gaseous or particulate compounds emitted through chimneys, ventilation systems, and other openings)
- Water (compounds released with any wastewater)
- Soil (solid waste to landfills, sewage sludge to agricultural soils)

Resource consumption is divided according to whether the resource consumed is renewable or nonrenewable.

### 9.4.2 Quality of Data

The quality of the data collected in the inventory is crucial to the outcome of the LCA, and therefore the choice of information sources should be considered carefully. In general, the best data are data that have been measured or calculated based on mass balances for the specific process performed on the actual equipment that is relevant for the product system. This holds true also for LCAs conducted with a broader scope, such as describing an average situation for a product within a whole region. In this case, the study must gather primary data from installations representing the different levels of performance represented in the region, so that a weighted value can be found.

The collection of this kind of specific data will, in general, be very time-consuming, and therefore the collection of process data should be guided by a sensitivity analysis focusing the effort on the data that have the greatest influence on the overall outcome of the life cycle assessment.

Very often it will not be possible to get specific data for some processes, and it will then be necessary to use data that have been extrapolated from the same or similar process technology elsewhere or even from other types of technology.

In some situations it may not even be possible to obtain *quantitative data* for a process, and a qualitative description may have to be included in the inventory. *Qualitative data* are very difficult to handle in the subsequent impact assessment step of the LCA. However, *absence of data* is worse.

An important point to keep in mind concerning data obtained from literature is that, due to the delay involved in writing and publishing, it will generally be at least five years old. For industrial processes in particular, this is a serious drawback, since the optimization and development of new, improved processes often proceeds rapidly. Consequently, the data on environmental releases and resource consumptions will be obsolete or will describe only the poorest functioning processes on the market. It is therefore important to ensure that the literature data cover technologies that are representative of those in the product system.

General *rules of thumb* concerning the quality of data for the inventory prescribe the use of:

- The most recent data
- Quality-assured and declared data
- Specific data whenever relevant and possible for both specific and general LCAs
- General or estimated data when sufficient and when specific data are not available
- Quantitative data when possible

Quality control of the data should be performed, e.g., by using mass and energy balances over given processes, or by checking the values against other sources in the literature.

The LCA-report should contain a *thorough documentation — process by process — of the sources of data used.* If possible, an estimate of the precision and variation of the data should also be provided. Use of low-quality data should be based on sensitivity analysis substantiating that the outcome of the LCA is not sensitive to these data.

### 9.4.3   Use of Computer Tools and Databases

Even for simple products, the life cycle can easily involve a large complexity of processes. As soon as electricity is used the complex process system behind electricity generation forms part of the product system, and even life cycles for simple materials such as metals tend to involve rather complex process systems. Furthermore, important data will often be of a confidential character. Obviously, for LCAs conducted for internal use this is not a problem. However, for LCAs meant for external uses it may cause serious difficulties, limiting the public access to, and the possibilities of critical review of, the results of the LCA.

Therefore, the existence of databases with unit process data for the most common processes and materials where the use of general data is justified is today a prerequisite for performance and public review of life cycle assessments. Publicly available LCA databases serve a double purpose of facilitating public use and control of the LCA tool, and of making the results of life cycle assessments more reproducible. They are therefore of benefit to both industry and the public. Fortunately, several comprehensive and good LCA databases exist today. (Frischknecht et al., 1996; Frees and Pedersen, 1996). Hemming (1995) gives a detailed overview of existing life cycle inventory data sources.

The later years have also seen the development of a wide range of different computerized tools for assisting the performance of life cycle assessment. The tools in general contain a module for modeling of the product system and generation of the inventory. The inventory analysis may be assisted by the incorporation of a database in the tools, and some tools also contain a module for the impact assessment assisting the interpretation of the inventory data. References are given in the reference list to some of the important tools in the market today. Weidema (1997) gives an excellent discussion of the virtues of the most prominent tools.

## 9.5   Impact Assessment

The inventory provides information about exchanges between all the processes of the product system and the environment. If the inventory has been made thoroughly it will in general contain a very large number of substance emissions and input of different resources. Some of these exchanges are environmentally significant and even small amounts can be important. Others are of no significance.

It is the task of the impact assessment phase to interpret the inventory results into potential effects on what is referred to as the "protection areas" or "safeguard subjects" of the LCA, i.e., the entities that we want to protect by performing and using the LCA. Today, there is general acceptance that the protection areas of life cycle assessment are (Udo de Haes, 1996):

- Human health
- Ecosystem health
- The resource base

   For the life cycle assessment to be able to support decisions, the data in the inventory must be interpreted. The interpretation must be based on the available background knowledge of the environment, resources, and working environment, and it must show which of the exchanges are significant through their impacts and potential effects on the protection areas, and how great their contributions can be.

For the environmental part of the impact assessment the ambition is thus to interpret the emissions into their potential effects on the protection areas through the environmental causality web illustrated in Figure 9.5.

The interpretation performed in the assessment phase of LCA normally proceeds through four steps, which will be reviewed below:

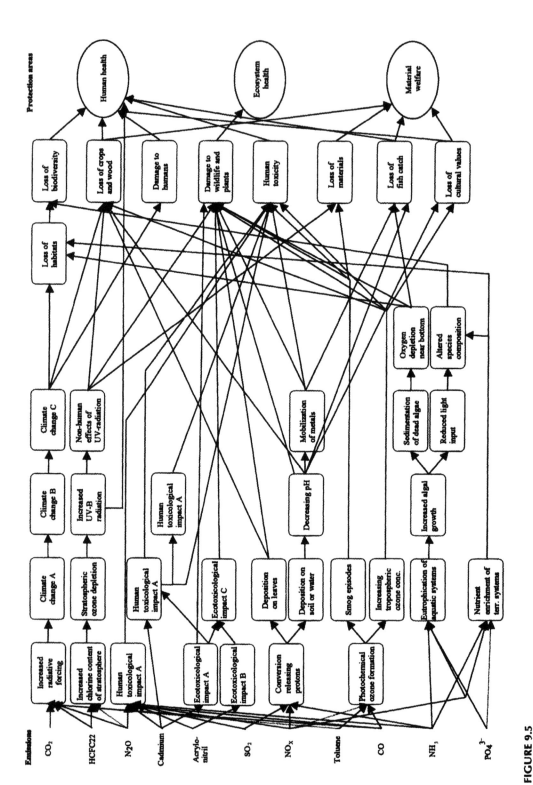

**FIGURE 9.5**
Causality web or cause–impact network for environmental emissions (not exhaustive).

- **Classification**, where the impact categories are defined and the exchanges from the inventory are assigned to impact categories reflecting their ability to contribute to different problem areas. (*"What is the problem for this environmental exchange?"*)

- **Characterization**, where the contribution(s) of each exchange is modeled quantitatively and the contributions aggregated within each impact category converting the classified inventory into a profile of environmental impact potentials, resource consumptions, and possibly working environment impact potentials. (*"How big is the problem?"*)

- **Normalization**, where the different impact potentials and resource consumptions are related to a common reference in order to facilitate comparisons across impact categories. (*"Is that much?"*)

- **Weighting**, where weights are assigned to the different impact categories and resource consumptions reflecting the relative importance they are assigned in this study in accordance with the goal of the study. (*"Is it important?"*)

A large part of the method development activities during the 1990s have dealt with the life cycle impact assessment phase. The review given here is based primarily on the EDIP method (Wenzel et al., 1997; Hauschild and Wenzel, 1998), and there may be minor deviations from other methodologies, particularly for the assessment of toxicological properties of the hazardous substances in the inventory, where the methodology is still rather immature and consensus lies well ahead. Important other impact assessment methodologies are those developed by Dutch researchers led by the group at the Leiden University (Heijungs et al., 1992; Guinée et al., 1996) and by Swiss researchers at the Federal Institute of Technology at Lausanne (Jolliet and Crettaz, 1997).

### 9.5.1 Classification

The first step of the life cycle impact assessment involves the definition of the impact categories that were chosen for the assessment in the scope definition phase. Normally most or all of the impact categories included in the study have already been defined earlier in other contexts and can simply be adapted for the present study.

The impact categories should be defined based on recognized environmental science, and double counting should be avoided, i.e., the impact categories should be defined at the same level of the causality web or, if this is not possible, in such a way that there is no overlap between the sections of the web that are covered by the different impact categories.

The next step is to go through the inventory and for each emission identify the impact categories to which it has the potential to contribute. Some of the substances may contribute to more than one impact category (as an extreme example, $NO_x$ contributes to the impact categories "acidification," "nutrient enrichment or eutrophication," "photochemical ozone formation," and "human toxicity").

### 9.5.2 Characterization

Based on the outcome of the classification, the potential contributions from the emissions of the life cycle can now be calculated for all the impact categories. When we are talking about "potential contributions" and "impacts" or "impact potentials" rather than "effects" it is due to some inherent characteristics of the product system that is studied in life cycle assessment.

- The object of the study is the functional unit, and the inventory provides information about that share of the total environmental exchanges from the processes in the product system that is caused by one functional unit. In LCA we are thus not concerned with the total output and impact from the processes but a fraction that varies from process to process according to the number of functional units per time unit that draw on the process.

- Environmental effects arise as a consequence of the total load or impact on a given receptor or receiving environment. To assess the effects of a process thus requires knowledge about the concomitant emissions from other processes impacting the same receptor and the background concentration of relevant substances in the system. Given the complexity and the global nature of the product system, this type of information will at best be available for a few of the processes in the product system.

- Focusing on the life cycle of the product, the LCA aggregates impacts occurring in different places and even at different times given the fact that many years may pass between the extraction of the raw materials and the final emissions from waste disposal in the life cycle of the product.

As a consequence, life cycle impact assessment aims at determining the environmental impact potentials of the emissions in the inventory, not real effects. Incidentally this also implies that life cycle impact assessment models are very difficult to validate because product systems are fictitious entities that we cannot monitor in the real world. These aspects of life cycle impact assessment are discussed in greater detail in Potting and Hauschild (1997a and b).

For all substances identified as contributing to a given impact category the contribution (the impact potential) is modeled quantitatively in the characterization step. The modeling as such is normally done in a different context outside the LCA study and introduced into the study through a compilation of characterization factors, which for each substance expresses the potential contribution to the impact relative to that of a reference substance chosen specifically for that impact category. The characterization is thus based on equivalence modeling expressing how many grams of the reference substance are required to give an impact contribution equivalent to that from one gram of the substance. For the impact category global warming, the global warming potential, GWP, of a substance is chosen as characterization factor, and the impact modeling is done under the auspices of the Intergovernmental Panel on Climate Change, IPCC, and representing the integrated radiative forcing of the substance. As reference substance for global warming carbon dioxide, $CO_2$, is chosen, and the characterization factors express the substances' potential impacts as grams of $CO_2$ equivalent per gram of substance. When nitrous oxide has a characterization factor of 310 integrating over a time horizon of 100 years, it means that over that time span, an emission of 1 gram of nitrous oxide contributes as much to global warming as the emission of 310 g $CO_2$. Table 9.1 shows equivalence factors for some greenhouse gases contributing to global warming.

Generally expressed, if the emission of a substance (i) has the magnitude $Q_i$ and if the substance's characterization factor for the impact category (j) is $EF(j)_i$, the emission's potential contribution $EP(j)_i$ to the impact category is calculated as:

$$EP(j)_i = Q_i \cdot EF(j)_i$$

Since characterization expresses all contributions to an impact category relative to the same reference substance, the total contribution seen over the entire product system can be calculated simply by adding the contributions of all individual emissions:

$$EP(j) = \Sigma EP(j)_i = \Sigma(Q_i \cdot EF(j)_i)$$

**TABLE 9.1**

For characterization of contributions to the impact category global warming the global warming potentials developed by IPCC are normally adopted as equivalence factors in LCA. Often a time horizon of 100 years is chosen.

| Substance | Formula | Global warming | | |
|---|---|---|---|---|
| | | 20 years | 100 years | 500 years |
| Carbon dioxide | $CO_2$ | 1 | 1 | 1 |
| Methane | $CH_4$ | 56 | 21 | 6.5 |
| Nitrous oxide | $N_2O$ | 280 | 310 | 170 |
| Methylene chloride | $CH_2Cl_2$ | 31 | 9 | 3 |
| HFC-32 | $CH_2F_2$ | 2,100 | 650 | 200 |
| HFC-134 | $CH_2F_4$ | 2,900 | 1,000 | 310 |
| Perfluoromethane | $CF_4$ | 4,400 | 6,500 | 10,000 |
| Perfluorobutane | $C_4F_{10}$ | 4,800 | 7,000 | 10,100 |
| Sulfur hexafluoride | $SF_6$ | 16,300 | 23,900 | 34,900 |

Source: Houghton, J.T., Meira Filho, L.G., Callander, B.A., Harris, N., Kattenberg, A., and Maskell, K., 1996, *Climate Change 1995. The Science of Climate Change*, Cambridge University Press, Cambridge. With permission.

## Choice of Characterization Modeling

As mentioned earlier, the choice of the model used for expressing contributions to an impact category should normally be based on science and be generally accepted. The choice of GWP as characterization factor for global warming is a good example of this approach, but for several of the other impact categories there is no international scientific consensus panel deriving equivalence factors, and the characterization modeling must then be performed within the study. Such models are derived for the most common LCA impact categories by Hauschild and Wenzel (1998) (the EDIP method) and a broader review of different approaches is given by Lindfors et al. (1995).

## Site-Specific Considerations

One of the features of the environmental assessment of products is that the inclusion of site-specific conditions for most of the processes in the product system must be of a very general nature due to the complexity and worldwide nature of the system. In fact, in many cases there is no site-specific information at all. As a consequence, it is in general not possible to include exact information on site-specific conditions, and as a consequence of this, it is also not possible to make a specific assessment of the *exposure* of those elements of the environment (the target systems) which are sensitive to the substances emitted from the product system. State-of-the-art in characterization modeling is thus for the nonglobal impact categories (i.e., all except global warming and stratospheric ozone depletion) to refrain from modeling the exposure of the target system.

Experiences with environmental assessment of products have, however, revealed a substantial need to attempt to *approximate* the LCIA so that an assessment of exposure is included. For emissions from certain processes it is quite obviously unreasonable to calculate a potential contribution to an impact category, if it can be said with certainty that the emission cannot trigger an effect where the process is occurring. This can, for example, be the case for human toxicity, when toxic but short-lived substances are emitted at a place where humans are not exposed to them. Or it can be the case for emission of $SO_2$ from ocean-going shipping, where it can be said with certainty that this $SO_2$ will never be able to contribute to acidification because it is absorbed by the sea, which, because of its enormous volume, is not sensitive to acidification. The issue of including spatial information in characterization modeling is treated in detail by Potting and Hauschild (1997a and b).

As one of the first methods, the EDIP method has been prepared for this kind of site characterization by admitting inclusion of a site factor which, based on the available knowledge of site-specific conditions, allows inclusion of generalized exposure information in the calculation of the potential environmental impact.

$$EP(j) = \Sigma\Sigma EP(j)_{p,i} = \Sigma\Sigma\ (Q_{i,p} \cdot EF(j)_i \cdot SF(j)_p)$$

where i is the individual substance, p is the individual process, and j is the impact category in question, and SF is the site factor representing the relevant site-dependent source and target information.

As is evident in the formula, calculation of the potential environmental impact must occur separately for each process, *before* the processes can be combined in the aggregation for the product system. When site-specific aspects are included, the emissions cannot, therefore, be combined for the processes before the calculation of impact, as is otherwise done.

If no information on site-specific conditions is available, SF is assigned the value 1. This corresponds to the normal characterization practice for LCA. It means that the conservative viewpoint is adopted, that the full potential impact will trigger real effects in the environment unless different information is available. The magnitude of the site factor can be regarded as the *probability* that real environmental effects will occur as a consequence of the type of emission in question. It will still not be possible to predict real effects in specific recipients, and it is thus still a matter of *potentials* for environmental effects. But by use of a generalized exposure assessment in the form of a site factor, the LCA provides a more accurate picture of the effects to which the product may contribute.

A different approach to inclusion of spatial aspects in characterization modeling for non-global impact categories is development of new characterization factors integrating fate, exposure, and effect of the substance where this is possible. Potting et al. (1997) presents such integrated characterization factors for the impact category acidification.

### Interpretation of Characterization Results

The calculated scores for each of the impact categories together constitute the characterized environmental profile of the product. In comparing the potential impacts from alternative products, one alternative will sometimes have a lower contribution to all impact categories than the other. If the same is the case for all resource consumptions and working environmental impacts the characterization has fulfilled a significant objective, as it is now clear which alternative is preferable from an environmental perspective. On the other hand, if the result is so clear, it will rarely come as a surprise to the person doing the LCA.

Most often, one alternative will have the lowest impact potentials for some categories, and a different alternative lowest potentials for other. In this type of situation, a comparison of impact potentials across categories is necessary, allowing an overall evaluation involving the assessment of the various impact potentials and resource consumptions relative to each other. Are any of them greater and more serious than others?

Such a mutual comparison of the magnitudes of the potentials is unreasonable without further consideration. Concerning often very different environmental problems, it is really a comparison of apples and pears. Furthermore, and more fundamentally, the quantities are calculated and expressed in different units. The global warming potential as g $CO_2$ equivalents, the ozone formation potential as $C_2H_4$ equivalents, and these values are not comparable. The first step in the comparison across impact categories is therefore to relate all the potential impacts to a common scale, a common reference, expressing them in the same units and thus revealing which impacts are large and which are small relative to this reference impact. This step is called normalization.

### 9.5.3  Normalization

Normalization, the scaling of all impact potentials and resource consumptions using a common reference, has two purposes:

- To provide an impression of the relative magnitudes of the potential impacts and resource consumptions
- To present the results in a form that is suitable for the final weighting and decision-making

For the latter purpose it is important that the normalization references represent an impact of known magnitude and contribution to effects on the environment, working environment and depletion of reserves.

The potential impacts and resource consumptions determined for the product systems are compared with a reference impact which is common for all impact categories, and the consequences of which for the environment, resources, and possibly working environment are known. In this way an impression is gained of which potential impacts are large and which are small, seen in relation to the known reference impact.

### *Normalization References*

As reference for normalization it is common to use a measure of the current impacts from society. The normalization thus expresses the different impact potentials of the product as fractions of society's total environmental impact. Even if all of the future consequences of society's impact on the environment and the working environment are not known, we still have an idea of how serious the situation is for each individual impact category, and this knowledge is important for the subsequent weighting.

Normalization references for the potential environmental impacts are determined on the basis of an inventory of the reference year for all of society's emissions which can contribute to the impact category. When the magnitude of the background emissions from society is determined, they are converted to environmental impact potentials, as described in Section 9.5.2 on characterization.

The impact of society's activities changes with time, for example as a consequence of changes in consumption and living standard and as a consequence of targeted initiatives against the worst impacts. It is therefore necessary to choose the same reference year for all impact categories within each of the three categories, environment, resources and working environment, to ensure that there is a common scale for all impact categories.

The normalization consists in dividing the impact potentials or resource consumptions by the corresponding normalization references. The normalized environmental impact potentials, NEP, are thus calculated as:

$$NEP(j) = \frac{EP(j)}{ER(j) \cdot T}$$

If the functional unit defines the duration of the service as T years, the normalization reference is expressed as $T \cdot ER(j)$, where $ER(j)$ denotes the normalization reference for one year for the impact category.

### Geographical Scale

Some of the environmental problems are of global nature, i.e., their environmental impact is the same regardless of where on the surface of the Earth the emission occurs. (They are generally caused by substances that have a lifetime sufficiently long to ensure a uniform global distribution.) Others are of a regional or local character. Emissions of greenhouse gases thus contribute to global warming, irrespective of where in the world they occur. But the Far Eastern or North American emissions of substances which contribute to acidification have no influence on the degree of acidification of forests and lakes observable in Europe today.

This is an important characteristic to consider in the weighting that follows the normalization step and for which the normalization serves as a preparation. The normalization references shall correspond to an impact of known magnitude and the impacts should thus be calculated for that area which actually contributes to the current state of the environment. In the EDIP method for life cycle impact assessment the normalization references are therefore chosen to reflect the scale of the different impact categories. For those impact categories which work on a global scale, the EDIP method uses the total global impact as normalization reference. For the other impacts on a regional or local scale the normalization is based on regional impacts in Europe or Denmark (Hauschild and Wenzel, 1998).

### The Person Equivalent

The global impact will always be much greater than the impact of the emissions from a particular region because the global impacts are the result of the activity of many more people than the impacts from the region. Use of global impacts as normalization reference for the global impact categories and, e.g., European impacts for the regional impact categories will thus introduce an imbalance in the normalization, causing global impacts from the product system to appear much less than the other impacts, simply because they are compared with the activity of the population of the entire world, while the regional impacts of the product are compared with the activity within the European population. To correct this imbalance and ensure that the set of normalization references constitutes a common scale for all impact categories, irrespective of whether they are global or regional, society's background impact over the course of one year is normalized with the population within the area for which the background impact is calculated. This gives the normalization references the unit "impact *per person* per year," or "person equivalent" for each individual impact category.

> When the impact potentials are normalized using the EDIP normalization reference, they are expressed in person equivalents (PE), i.e., fractions of the background impact deriving from the average person.

The approach is the same for normalization of the resource consumptions. The normalization references for use of nonrenewable resources are based on total global consumption of the resource, because resources are traded at one common world market. The region where the resource originates is therefore uninteresting from a resource point of view. For renewable resources, however, it may be decisive in the subsequent weighting whether consumption exceeds renewal or regeneration of the resource. This can vary from region to region and from local area to local area. In the EDIP method the normalization references

| Category | Unit | 1 refrigerator per year | 1 person per year | Per thousand |
|---|---|---|---|---|
| **Environmental impact potentials** | | | | |
| Global warming | g $CO_2$-eq/year | 286,000 | 8,700,000 | 32 |
| Ozone depletion | g $CFC11$-eq/year | 48 | 198 | 240 |
| Acidification | g $SO_2$-eq/year | 615 | 138,000 | 4.6 |
| Nutrient enrichment | g $NO_3^-$-eq/year | 400 | 264,000 | 15 |
| Hazardous waste | g/year | 750 | 20,700 | 28 |
| **Resource consumption** | | | | |
| Crude oil | g/year | 2,810 | 591,000 | 48 |
| Coal | g/year | 24,000 | 573,000 | 42 |
| Aluminium | g/year | 36 | 3,370 | 11 |
| Nickel | g/year | 0.85 | 177 | 48 |
| **Impact potentials on the working environment** | | | | |
| Noise that induces hearing impairments | hours/year | 0 0982 | 514 | 0 19 |
| Allergenic substances | hours/year | 0 0378 | 62 | 0 61 |
| Carcinogenic substances | hours/year | 0.00831 | 47 | 0 18 |

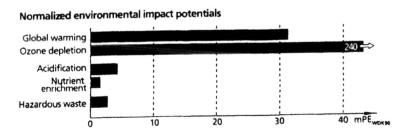

**Normalized environmental impact potentials**

**FIGURE 9.6**

Relating the product's impact potentials and resource consumption to those of an average person (an expression of the background impact from society's activities), normalization brings environmental impact potentials, resource consumption, and potential impacts on the working environment on the same scale. This allows contributions to different environmental impact categories to be expressed in the same unit — the milli person equivalent (mPE) — and presented in the same bar diagram as shown for some of the impacts from the example — a refrigerator. The index "WDK90" of the person equivalent unit refers to the fact that for the global impact categories the person equivalent is determined using background emissions for the world (W) in the reference year 1990, while for the nonglobal categories, regional background emissions (in this case for Denmark, DK) are used. (From Wenzel, H., Hauschild, M., and Alting, L., 1997, *Environmental Assessment of Products, Volume 1, Methodology, Tools and Case Studies in Product Development*, Chapman and Hall, London. With permission.)

for renewable resources are therefore based on the total consumption per person in the relevant area.

On normalization, all potential impacts and resource consumptions thus assume the same unit, and it is possible to compare their magnitudes. At the same time, the normalized potential impacts are expressed in a comprehensible unit as they can be viewed relative to one's own average contribution to the impact.

The person equivalent is illustrated in Figure 9.6. The table on top shows some average annual resource consumptions and potential impacts for the environment and working environment for the product system of a refrigerator. The refrigerator's potential impacts are seen in relation to the background impact from an average person. The refrigerator's potential impacts and resource consumption can thus be expressed in parts per thousand of a person's average impact for the various impact categories or milli-person equivalents, mPE, and shown in the same bar graph as illustrated for the environmental impact potentials in the lower part of the figure.

### Use of Normalized Potential Impacts and Resource Consumptions

On normalization, the potential impacts and resource consumptions are expressed on a common scale, e.g., as person equivalents, and it is now possible to compare:

- The impacts from alternative products, from entirely different products, and in effect from any activity
- The relative magnitude of the contributions to the individual impact categories or resource consumptions within each of the domains: environment, resources, and working environment

With due regard for the difference between the impact categories within the three domains, it is also possible to compare the magnitude of the contributions between the domains in order to gain an impression of whether the product is particularly severe on the working environment or the environment, or whether it is the resource consumption which has the greatest impact relative to society's total activities. If a comparison between impact categories or across domains produces results in conflict with what would be expected, it can mean that significant exchanges have not been included for one of more of the categories.

In addition, when using the EDIP normalization procedure,

> on normalization, the magnitudes of the potential impacts and the resource consumptions are expressed in a unit to which it is simple to relate, namely fractions of the annual impact from an average person.

Even if normalization reveals which impacts are large and which are small relative to society's background impact, there will always be situations where judgment is required in comparison of alternative products. This is the case in trade-off situations where one alternative has the lowest normalized potentials for certain impact categories or resource consumptions, while another alternative contributes least to others. In such a situation, the various impact categories and resource consumptions must be weighed relative to one another before they can be compared. Are any of them more serious than others, and how much more serious?

This weighing must aim at expressing the relative seriousness within each of the three domains: environment, resources, and working environment.

### 9.5.4 Weighting

Normalization assists the comparison of different impact potentials by relating them to society's background impact. However, even if the contributions to two different impact categories are equally large on normalization, this does not automatically mean that the impact potentials are equally serious. To permit a direct comparison of the different impact potentials, an assessment must first be made of the seriousness of the impact categories relative to one another.

The mutual seriousness of impact categories or resource consumptions is expressed in a set of weighting factors with one factor per impact category or resource consumption allowing the weighting to be performed by multiplying the normalized impact potential or resource consumption, by this weighting factor:

$$WP(j) = WF(j) \cdot NP(j)$$

where WP(j) is the weighted impact potential or resource consumption, WF(j) the weighting factor for impact category or resource j, and NP the normalized impact potential or resource consumption.

### 9.5.4.1 Weighting of Potential Environmental Impacts

The weighting factor for an environmental impact must reflect the seriousness of the impact relative to the other environmental impact categories that are considered in the LCA. By seriousness is meant its possible consequences for the protection areas or safeguard subjects of the LCA:

- Human health
- Ecosystem health
- Resource base

The weighting factors assigned to the individual impact categories thus should ideally reflect the ability of the impact indicator to affect these three protection areas and the relative importance that is assigned to the protection areas. The determination of the weighting factors should therefore involve an analysis of the web of cause–impact relationships between emissions and protection areas as exemplified earlier in Figure 9.5.

The analysis of the causality web should involve aspects like:

- What level of effects can be anticipated from the impact (extinction of species, death of individuals, impairment of function of species/ecosystems)?
- How great an area will be affected by the damage?
- How certain are we about the causal relationships between the emissions and these effects?
- How far is the current impact status from critical threshold values for the effect in those areas which are affected by the emission?
- When will these effects be felt?
- Will the environmental damage be reversible, and if so, for how long will the effects persevere after the impact ceases?

Weighting inherently involves expression of values. It is thus not possible on a purely objective and scientific basis alone to develop weighting factors. The relative importance assigned to the three protection areas, to future effects relative to current effects, to the uncertainties disclosed by the analysis of the environmental causality web in Figure 9.5 depends on ethical values. If the different stakeholders in the decision process into which the LCA enters are to accept the results as valid, they must experience the weighting as fair and in agreement with their own ethical values.

Due to the many uncertainties and gaps in our current knowledge about environmental mechanisms, it has proven very difficult to derive weighting factors with a common acceptance based on an analysis of the environmental causality web.

Depending on the decision context into which the LCA will enter, other characteristics of a more political nature may also be considered. If the LCA is to be used by a company to rank an initiative in relation to a particular market, questions involving attitudes could, for example, be:

- What is the perception of the seriousness of the environmental effect among potential buyers of the product or others among the company's interested parties?
- What is the perception among various opinion leaders in the community in general?

If the LCA is used by authorities, for example to set criteria for assigning an ecolabel to a product, it would be relevant to consider existing regulation, political action plans, conventions, and international agreements addressing the most important impacts for the product type.

### Weighting on the Basis of Political Environmental Targets

The EDIP program developed weighting factors based on existing environmental policy targets as an expression of the importance that society assigns to the different problem areas. These weighting factors are an expression of the official environmental priorities and therefore have their relevance in studies where the regulatory bodies are among the stakeholders. In several industrialized countries a product-oriented environmental policy is being defined as a central part of the environmental administration, and for environmental assessment of products in this context the policy target-derived weighting factors seem appropriate. Apart from this application of the policy target-derived weighting factors, they may serve as proxies for weighting factors based on a detailed environmental analysis. As a rule, environmental research has continued on an impact for a very long time before action plans are initiated or reduction targets adopted. Those types of policy measures thus have a significant scientific content.

When targets are set for how much society's environmental impact is to be reduced, this is done on the basis of considerations of how serious the consequences of the impact can be and the costs which will be associated with reducing them. The considerations include such aspects as:

- What damage to the environment can be observed today as a consequence of the impact?
- What damage to the environment can be expected as a consequence of the impact, and what environmental consequences can result in the short and the long term?
- What costs will this damage impose on society?

But also aspects which have little to do with the extent of the environmental impacts:

- What technological possibilities are available for preventing and repairing the damage?
- Is the public aware of the environmental effect?
- How will the planned initiative against the impact affect the national and the international economies and employment?

For the individual environmental impact, the political setting of targets for reduction thus implies a balancing of scientific and political considerations. No conscious balancing of the seriousness of this environmental impact is made relative to the seriousness of

**TABLE 9.2**

Weighting Factors Based on Danish Environmental Policy Targets
Inter- or Extrapolated to Represent the Period 1990–2000

| Impact Category | Weighting Factor |
|---|---|
| Global warming | 1.3 |
| Ozone depletion | 23 |
| Photochemical ozone formation | 1.2 |
| Acidification | 1.3 |
| Nutrient enrichment | 1.2 |
|    N-equivalents | 1.3 |
|    P-equivalents | 1.0 |
| Persistent toxicity | 2.5 |
| Human toxicity | 2.8 |
| Ecotoxicity | 2.3 |

Source: From Wenzel, H., Hauschild, M., and Alting, L., 1997, *Environmental Assessment of Products, Volume 1, Methodology, Tools and Case Studies in Product Development*, Chapman and Hall, London. With permission.

the other impacts to which the environment is exposed. The targets for reductions are, however, fixed within society's total economic framework for environmental improvements, and the initiative regarding individual substances and groups of substances is therefore indirectly ranked in relation to the total environmental initiative. On this basis, the political setting of reduction targets can be considered a result of a decision-making process similar to that which underlies the determination of weighting factors for the environmental impacts.

Political reduction targets are normally established for individual substances or groups of substances and not for society's total contributions to environmental impacts. For example, many countries have set a target for reduction of the national emissions of $CO_2$, which is the most significant greenhouse gas, but not for a reduction in the total national contribution to global warming, which is also attributable to substances other than $CO_2$. In the derivation of weighting factors, the reduction targets for individual substances must therefore be translated into reduction targets for environmental impacts in the same way that the inventory of environmental exchanges is translated into environmental impact potentials in the characterization step.

As a rule, the reduction targets are formulated such that society's emissions of a substance or a group of substances in the selected target year may amount at most to a certain percentage of the emissions in a reference year. But reference year and target year vary for the various substances and groups of substances, depending on the time when the reduction targets are set, and also on the desirable and realistic time frame for achievement of the reductions. To give a uniform treatment of all environmental impact categories, in the EDIP method the targets for reductions are harmonized through linear inter- or extrapolation to apply to the same period for all impact categories before they are used as a basis for calculation of the weighting factors (Wenzel et al., 1997).

The weighting factors for environmental impacts is calculated as the current impact divided by the targeted impact for a common target year.

$$WF(j) = \frac{\text{Potential environmental impact of current emissions}}{\text{Potential environmental impact of emissions in target year}} = \frac{ER(j)_{\text{reference yr.}}}{ER(j)_{\text{target yr.}}}$$

The weighting factor is thus determined as the extent to which the normalization reference must be reduced in the target year to be in agreement with the efforts inherent in the reduction targets that are relevant for the environmental impact category in question. The more ambitious the reduction targets, the lower the targeted impacts and the higher the weighting factor.

Applying Danish reduction targets Table 9.2 shows the resulting weighting factors for the different environmental impact categories using 2000 as target year and 1990 as reference year for current emissions.

With this definition of weighting factor, the weighting can be seen a sort of normalization with the targeted emissions as normalization reference, as the weighted potential environmental impact WEP(j) is determined as:

$$WEP(j) = WF(j) \cdot NEP(j)$$

$$WEP(j) = \frac{ER(j)_{reference\ yr.} \cdot EP(j)}{ER(j)_{target\ yr.} \cdot ER(j)_{reference\ yr.} \cdot T}$$

$$WEP(j) = \frac{EP(j)}{ER(j)_{target\ yr.} \cdot T}$$

In other words, using these environmental weighting factors the environmental impact potential for the product is expressed as a percentage of the targeted person equivalent, PET, which can be expected in the year 2000 if society's plans for reduction are achieved.

The "environmental latitude" or "environmental space" is the environmental impact that is available on average to each person in a sustainable society. In the same way, the PET is the "environmental policy target latitude" for the target year, i.e., the impact which we, on average, may cause for each of the impact categories if the targets for reductions are to be fulfilled. The size of the environmental policy target latitude will gradually approximate the size of the environmental latitude as the environmental policy targets approximate the targets for sustainability.

The unit of the weighted impact potentials is thus immediately comprehensible. If one buys a car with a global warming potential of 500 mPET, half of one's "ration" for the contribution to global warming has been used for all of the time during which that car is used.

### 9.5.4.2   *Weighting of Resource Consumption*

Basically, the weighting of a consumption of a resource must reflect why the consumption of this resource is considered a problem. Although there seems to be a common perception that an important issue is the availability of the resource for future generations (a central element in the sustainability concept), there has not yet emerged a common approach to the weighting of resources. Several approaches use the supply horizon — the number of years for which current consumption of a nonrenewable resource can continue before all known reserves are exhausted. This type of weighting factor reflects how scarce the resource is relative to consumption of it (Wenzel et al., 1997).

A different approach weights the resource consumption with a measure of the energy required to restore the resource from the way it is used in the product. The energy measure may be pure energy or a measure of the exergy (energy available for doing work) involving also the entropy changes (Finnveden, 1994).

For renewable resources a different aspect exists. As long as they are not consumed faster than their rate of regeneration, there is no depletion and their consumption is generally not considered a problem in life cycle impact assessment. While consumption of the nonrenewable resources can be considered as a global impact (the impact on the resource base is the same regardless of where the resources are extracted), there is generally a strong regional dependence for the renewable resources. In the Scandinavian countries the standing forest biomass is increasing, and this resource is thus currently regenerated faster than it is used. For forests in large parts of the Earth's tropical regions, the opposite is the case and the forested area is reduced year by year.

If a renewable resource is consumed with greater speed than it is regenerated, it will in the long run be spoiled, and this should be reflected in the weighting of the resource consumption.

As an example of weighting of resource consumptions the EDIP method defines the weighting factor for resource consumption as the reciprocal of the supply horizon for the resource, calculated on the basis of consumption, the established reserves, and any rate of regeneration in the reference year.

$$WF(j) = \frac{1}{\text{Supply horizon for resource } (j)} = \frac{RR(j)_{\text{reference yr.}}}{\text{Known reserves of } (j)}$$

where $RR(j)_{\text{reference yr.}}$ is the total consumption of the resource in the chosen reference year. For nonrenewable resources, this definition of the weighting factor means that consumption, $R(j)$, of resource $(j)$ in the product system is compared against reserves of the resource in question at the weighting. The weighted resource consumption, $WR(j)$, is found as:

$$WR(j) = WF(j) \cdot NR(j)$$

$$WR(j) = \frac{RR(j)_{\text{reference yr}} \cdot R(j)}{\text{Known reserves of } (j) \cdot RR(j)_{\text{reference yr.}}}$$

$$WR(j) = \frac{R(j)}{\text{Known reserves of } (j)}$$

The weighted resource consumption is given the unit "person-reserve," PR, i.e., the proportion of known reserves per person in the world in 1990. These figures are also immediately comprehensible. For example, with a figure of 10 $mPR_{W90}$, it is possible to buy 100 products of the relevant type, and the "ration" of known reserves has thus been used for the entire future for all generations, i.e., also that portion of the known reserves which were otherwise available for one's children, grandchildren and subsequent generations.

It should be noted that the size of the reserves is not absolute, but depends to a high degree on the market price of the resource and on its strategic significance. The more expensive or important the resource, the more intensively it will be sought, the more inaccessible the places on Earth where it is sought (and the greater is the environmental impact per unit of resource extracted). The known reserves of crude oil have thus been more or less constant over the last few decades despite a continuing large consumption of the resource. In spite of this situation, the supply horizon defined on the basis of the size of known reserves is the best measure we have for the scarcity of the resource.

For renewable resources, the weighting factor is defined on the basis of the total consumption where the process is occurring. If the resource is consumed more rapidly than it is regenerated, a supply horizon can be defined as the number of years that will elapse before the resource is exhausted at the current ratio between present reserves and the difference between rate of consumption and rate of regeneration. The supply horizon for renewable resources is thus defined on the basis of actual consumption in the local areas where consumption is occurring. If the resource is not used faster than it is regenerated, the supply horizon is infinite and the weighting factor is therefore 0:

$$\text{Supply horizon} = \frac{\text{Known reserves}}{\text{Annual consumption} - \text{annual regeneration}}$$

## 9.6    Interpretation

In the interpretation phase of life cycle assessment the results are interpreted along the lines of the defined goal and in accordance with the limitations defined by the scope of the study. The outcome of the interpretation may be a conclusion of the study serving as a recommendation to the decision makers, who will normally weigh it against other decision criteria (like economic and social aspects). The interpretation may provide input to a further iteration, reviewing and possibly revising the scope of the study, the collection of data for the inventory, and the impact assessment.

In any case, a sensitivity analysis of the key parameters of the study (most important individual environmental exchanges or assumptions in the delimitation of the product system) is an indispensable part of the interpretation phase.

The sensitivity analysis identifies the key figures of the LCA — those model assumptions, processes, and environmental exchanges that have the greatest significance for the total result. The significance of the uncertainty in the key figures is examined by letting them vary within their estimated range and investigating how these will affect the total result of the LCA and the conclusions that can be drawn.

## 9.7    Reporting and Critical Review

As will be clear from the reading of this chapter, product systems are often very complex, and apart from this complexity, life cycle assessment involves a series of choices and assumptions that may render the outcome dubious or at least intransparent to people outside the study. It is therefore a requirement from both SETAC and ISO that there be a transparent and sufficiently detailed presentation of

- Results
- Data
- Methods used
- Assumptions made
- Inherent limitations of the study

to allow the reader to understand the complexities and trade-offs inherent in the study.

Apart from this it is recommended, and for some applications required, that a critical review of the study be performed by an independent third party. The review can be performed either after the study is finalized or during the study in interaction with the group doing the LCA study. The review report must be included in the reporting of the life cycle assessment.

---

# References

Association of Danish Electric Utilities, 1995, *Danish Electricity Supply Statistics 1994*, Rosenørns Alle 9, DK-1970 Frederiksberg C. (in Danish).

Boustead, I.,1993, *Polyethylene and Polypropylene, Eco-profiles of the European Plastics Industry, Report 3*, Association of Plastics Manufacturers in Europe (APME), Brussels.

Bundesamt für Umweltschutz (BUS), 1984, Ökobilanzen von Packstoffen, Schriftenreihe Umweltschutz no. 24, Bern.

Caspersen, N., 1995, Personal communication, Institute for Product Development, Technical University of Danmark, DK-2800 Lyngby, Denmark.

Danish Environmental Protection Agency, 1994, Note on quantities of waste in Denmark in 1993 compared with targets in the government's action plan for waste and recycling 1993–1997, Danish Environmental Protection Agency, Office for Industrial Waste, M356-0002, 1994, Copenhagen (in Danish).

Danish Environmental Protection Agency, 1996, *A Strengthened Product-Oriented Environmental Effort*, Copenhagen, (in Danish).

Danish Environmental Protection Agency, 1998, *Account of the Danish Environmental Protection Agency on the Product-Oriented Environmental Effort*, Copenhagen, (in Danish).

EC, 1999, Workshop on Integrated Product Policy, 8 December 1998, Final Report, European Commission, DG XI, Brussels.

Finnveden, G., 1994, Characterisation methods for depletion of energy and material resources, in Udo de Haes, H., Jensen, A.A., Klöpffer, W., and Lindfors, L.-G., Eds., *Integrating Impact Assessment into LCA*, Society of Environmental Toxicology and Chemistry — Europe, Brussels.

Fitzgerald, F., 1992, Energy use and management in British Steel plc, *Ironmaking and Steelmaking*, 19 (2), 98.

Franke, M., 1984, Umweltauswirkungen durch Getränkeverpackungen — Systematik zur Ermittlung der Umweltauswirkungen von komplexen Prozessen am Beispiel von Einweg-und Mehrweg-Getränkebehältern, Technische Universität Berlin/Institut für Technischen Umweltschutz, Berlin.

Frees, N. and Pedersen, M.A., EDIP unit process database (manual in Danish, database in Danish and English), Danish Environmental Protection Agency, Copenhagen.

Frischknecht, R., Hofstetter, P., Knoepfel, I. and Ménard, M., Eds., 1996, Life Cycle Inventories of Energy Systems (in German), Bundesamt für Energiewirtschaft (BEW), Projekt und Studienfonds der Elektrizitätswirtschaft (PSEL), Bern.

Fullana, P. and Puig, R., 1997, *Análisis del Ciclo de Vida*, Rubes Editorial S.L., Barcelona.

Guinée, J., Heijungs, R., van Oers, L., van de Meent, D., Vermeire, T., and Rikken, M., 1996, LCA impact assessment of toxic releases. Generic modelling of fate, exposure and effect for ecosystems and human beings with data for about 100 chemicals, Publikatiereeks produktenbeleid no. 1996/21, VROM, den Haag.

Guldbrandsen, T. and Nørgård, J.S.,1986, *Achieving Substantially Reduced Energy Consumption in European Type Refrigerators*, paper presented at the 37th Annual International Appliance Technical Conference, Purdue University, U.S.A., May 6-7, 1986.

Gydesen, A., Maimann, D., Pedersen, P.B., Hansen, M.K., Bruhn, B., and Bidstrup, C.,1990, *Cleaner Technology in the Area of Energy*. Environmental project 138, 1990, Danish Environmental Protection Agency, Copenhagen, (in Danish).

Hauschild, M. and Wenzel, H., 1998, *Environmental Assessment of Products, Volume 2, Scientific Background*. Chapman and Hall, London.

Heijungs, R., Guinée, J., Huppes, G., Lankreijer, R.M., Udo de Haes, H.A., Wegener Sleeswijk, A., Ansems, A.M.M., Eggels, P.G., van Duin, R., and de Goede, H.P., 1992, *Environmental Life Cycle Assessment of Products, Volume 1, Guide, Volume 2, Backgrounds*, CML, Leiden.

Hemming, C., 1995, *SPOLD Directory of Life Cycle Inventory Data Sources,* Society for the Promotion of LCA Development, Brussels.

Houghton, J.T., Meira Filho, L.G., Callander, B.A., Harris, N., Kattenberg, A., and Maskell, K., 1996, *Climate Change 1995. The Science of Climate Change,* Cambridge University Press, Cambridge,

Hunt, R.G., Franklin, W.E., Welch, R.O., Cross, J.A., and Woodall, A.E., 1974, *Resource and Environmental Profile Analysis of Nine Beverage Container Alternatives,* United States Environmental Protection Agency (U.S. EPA), Office of Solid Waste Management Programs, EPA/530/SW-91c), Washington, D.C.

ISO 14040, 1997, *Environmental Management — Life Cycle Assessment — Principles and Framework,* International Organization for Standardization.

Jolliet, O. and Crettaz, P., 1997, *Critical Surface-Time 95 (CST95). A Life Cycle Impact Assessment Methodology Including Exposure and Fate,* Federal Institute of Technology, IATE-HYDRAM, Lausanne.

Kindler, H. and Nickles, A., 1980, Energieaufwand zur Herstellung von Werkstoffen. Kunstoffe, 70. Carl Hauser Verlag, Munich.

Lindfors, L-G., Christiansen, K., Hoffmann, L., Virtanen, Y., Juntilla, V., Hanssen, O-J., Rønning, A., Ekvall, T., and Finnveden, G., 1995, *Nordic Guidelines on Life-Cycle Assessment,* (Chapter 7), Nord, 1995:20, Nordic Council of Ministers, Copenhagen.

Lundholm, M.P. and G. Sundström, 1985, Resource and environmental impact of Tetra Brik carton and refillable and non-refillable glass bottles, Tetra Brik Aseptic environmental profile, AB Tetra Pak, Malmö, Sweden.

Meadows, D. and Meadows, D., Eds., 1972: *The Limits to Growth: A Report for the Club of Rome's Project on the Predicament of Mankind,* Universe Publications, New York.

Mose, A.-M., Wenzel, H., and Hauschild, M., 1997, Gram: Refrigerators, Chapter 26 in Wenzel, H., Hauschild, M. and Alting, L., Eds., *Environmental Assessment of Products, Volume 1: Methodology, Tools and Case Studies in Product Development,* Chapman and Hall, London.

National Environmental Policy Plan, 1989, *To Choose or to Lose,* Ministry of Housing, Physical Planning and Environment, The Hague.

Nedermark, R., 1995, Personal communication, Bang & Olufsen A/S, Struer, Denmark.

Pedersen, B. and Christiansen, K., 1992, A meta-review on product life cycle assessment, in Lindfors, L.G., Ed., *Product Life Cycle Assessment — Principles and Methodology,* Nord, 1992, 9, Nordic Council of Ministers, Copenhagen.

Pedersen, J.,1995, Personal communication, Danish Energy Agency, Copenhagen.

Potting, J., Schöpp, W., Blok, K., and Hauschild, M., 1997, Site-dependent life-cycle assessment of acidification, *Journal of Industrial Ecology,* 2(2), 61.

Potting, J. and Hauschild, M., 1997a, Predicted environmental impact and expected occurrence of actual environmental impact. Part 1: The linear nature of environmental impact from emissions in life-cycle assessment, *Int. J. of LCA,* 2(3), 171.

Potting, J. and Hauschild, M., 1997b, Predicted environmental impact and expected occurrence of actual environmental impact. Part 2: Spatial differentiation in life-cycle assessment via the site-dependent characterisation of environmental impact from emissions, *Int. J. of LCA,* 2(4), 209.

Scholl, G., 1996, Sustainable product policy in Europe, *European Environment,* 6, 183.

Sustainability, 1993, The LCA Sourcebook. *A European Business Guide to Life-Cycle Assessment.* Sustainability Ltd., Society for the Promotion of LCA Development (SPOLD) and Business in the Environment, U.K.

Teoh, L., 1989, Electric arc furnace technology: Recent developments and future trends, *Ironmaking and Steelmaking,* 16, (5), 303.

Udo de Haes, H., Ed., 1996, *Towards a Methodology for Life Cycle Impact Assessment,* Society for Environmental Toxicology and Chemistry — Europe, Brussels.

Weidema, B.P., 1997, *Environmental Assessment of Products: A Textbook on Life Cycle Assessment,* Finnish Association of Graduate Engineers TEK, Helsinki.

Wenzel, H., Damborg, A., and Jacobsen, B.N., 1990, *Danish Emissions of Industrial Wastewater,* Environmental Project No. 153, Danish Environmental Protection Agency, Copenhagen, (in Danish with English summary).

Wenzel, H., Hauschild, M., and Alting, L., 1997, *Environmental Assessment of Products, Volume 1, Methodology, Tools and Case Studies in Product Development*, Chapman and Hall, London.

---

## LCA Software

*Simapro 4.0* from the Dutch consultancy company Pré: Database and modules for inventory analysis and impact assessment. Probably the most widespread LCA-software today (1998). Pré Consultants, Plotterweg 12, 3821 BB Amersfort, Netherlands.

*LCA Inventory Tool* developed by the Swedish consultancy company CIT: Modeling of product system and inventory analysis. Chalmers Industripark, 41288 Gothemburg, Sweden.

*KCL ECO* developed by the Finnish Pulp and Paper Research Institute: Modeling of product system and inventory analysis. Database module (KCL ECODATA) can be bought separately. KCL, Tekniikantie 2, P.O.Box 70, 02151 Espoo, Finland.

*TEAM* developed by the French consultancy company Ecobilan: Modeling of product system and inventory analysis. Comprehensive database modules can be bought separately. Ecobilan, 13-15 Rue Buffon, 75005 Paris, France.

*EDIP PC Tool* developed by the Institute for Product Development for the Danish EPA. Supports the EDIP methodology and contains database and modules for inventory analysis and impact assessment. Miljøbutikken, Læderstræde 1-3, DK-1201, Copenhagen K, Denmark.

# 10

## Environmental Risk Assessment

**S.E. Jørgensen and Bent Halling Sørensen**

## CONTENTS

## 10.1   What Is ERA?

Environmental risk assessment, ERA, may be defined as the process of assigning magnitudes and probabilities to the adverse effects of human activities. The process involves identification of hazards, such as the release of toxic chemicals to the environment, by quantification of the relationship between an activity associated with an emission to the environment and its effects. The entire ecological hierarchy is considered in this context, which implies that the effects on the cellular level, on the organism level, on the population level, on the ecosystem level, and for the entire ecosphere should be considered.

The application of environmental risk assessment is rooted in the recognition that:

1. The cost of elimination of all environmental effects is impossibly high
2. The decision in practical environmental management must always be made on the basis of incomplete information

We use about 100,000 chemicals in such amounts that they may threaten the environment, but we know only about 1% of what we need to know to be able to make a proper and complete environmental risk assessment of these chemicals. This chapter will give an overview of available estimation methods which can be recommended to apply if we cannot find information about properties of chemical compounds in the literature.

ERA is related to environmental impact assessment, EIA, which attempts to assess the impact of a human activity. EIA is predictive, comparative, and concerned with all possible effects on the environment, including secondary and tertiary (indirect) effects, while ERA attempts to assess the probability of a given (defined) adverse effect as result of a considered human activity.

Both ERA and EIA use models to find the expected environmental concentration (EEC), which is translated into impacts for EIA and into risks of specific effects for ERA. The development of models is not covered in this volume, although the applicability of different classes of models has been mentioned in Chapter 2 and will be further discussed in this chapter. Development of the ecotoxicological models that are applicable in assessment of environmental risks is treated in detail in Jørgensen (1994).

Legislation and regulation of domestic and industrial chemicals with respect to the protection of the environment has been implemented in Europe and North America for decades. Both regions distinguish between existing chemicals and introduction of new substances. For example, for existing chemicals the European Union requires, according to Council Regulation 793/93, an assessment of risk to man and environment of priority substances by principles given in Commission Regulation 1488/94. An informal priority setting (IPS) is used for selecting chemicals among the 100,000 listed in the *European Inventory of Existing Commercial Chemical Substances*. The purpose of IPS is to select chemicals for a detailed risk assessment among the EEC high production volume compounds, i.e., >1000 t/y (about 2000 chemicals). Data necessary for the IPS and an initial hazard assessment is called Hedset and covers issues as environmental exposure, environmental effects, exposure of man, and human health effects.

The risk assessment of newly notified substances is in EU based on the data submitted according to Directive 67/548/EEC. The directive provides a scheme of step-wise procedures which exist in North America and Europe approximately as presented below. Tests are often required to provide the data needed for the ERA.

At the UNCED meeting in Rio de Janeiro in 1992 on the Environment and Sustainable Development, it was decided to create an Intergovernmental Forum on Chemical Safety (IGFCS, Chapter 19 of Agenda 21). The primary task is to stimulate and coordinate global harmonization in the field of chemical safety, covering the following principal themes: assessment of chemical risks, global harmonization of classification and labeling, information exchange, risk reduction programs, and capacity building in chemicals management (Chichilnisky et al., 1997).

The uncertainty plays an important role in risk assessment (Suter, 1993). Risk is the probability that a specified harmful effect will occur or, in the case of a graded effect, the relationship between the magnitude of the effect and its probability of occurrence.

Risk assessment has emphasized risks to human health and has, to a certain extent, ignored ecological effects. It has, however, been acknowledged that some chemicals that have no or only little risk to human health cause severe effects to aquatic organisms, for instance. Examples are chlorine, ammonia, and certain pesticides. A up-to-date risk assessment therefore comprises considerations of the entire ecological hierarchy which is the ecologist's view of the world in terms of level of organization. Organisms interact directly with the environment, and it is organisms that are exposed to toxic chemicals. The species-sensitivity distribution is therefore more ecologically credible (Calow, 1998). The reproducing population is the smallest meaningful level in an ecological sense. However, populations do not exist in a vacuum, but require a community of other organisms of which the population is a part. The community occupies a physical environment with which it forms an ecosystem.

Moreover, both the various adverse effects and the ecological hierarchy have different scales in time and space which must be included in a proper environmental risk assessment (see Figure 10.1). For example, oil spills occur at a spatial scale similar to those of populations, but they are briefer than population processes. Therefore, a risk assessment of an oil spill requires considerations of reproduction and recolonization that occur on a longer time scale and that determine the magnitude of the population response and its significance to natural population variance.

Uncertainties in risk assessment are most commonly taken into account by application of safety factors. Uncertainties have three basic causes:

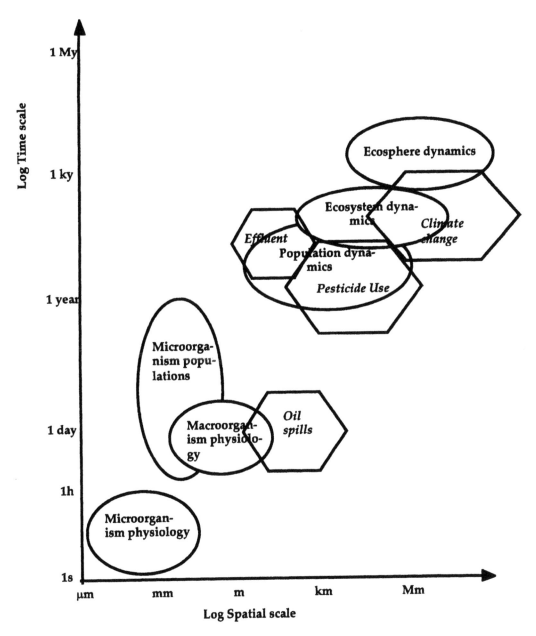

**FIGURE 10.1**
The spatial and time scale for various hazards (hexagons, italic) and for the various levels of the ecological hierarchy (circles, nonitalic).

1. The inherent randomness of the world (stochasticity)
2. Errors in execution of assessment
3. Imperfect or incomplete knowledge

The inherent randomness refers to uncertainty that can be described and estimated but cannot be reduced because it is characteristic of the system. The meteorological factors such as rainfall, temperature, and wind are effectively stochastic at levels of interest for risk

**TABLE 10.1**

Selection of Assessment Factors to Derived PNEC (see also step 3 of the presented procedure below)

| Data Quantity and Quality | Assessment Factor |
|---|---|
| At least one short-term $LC_{50}$ from each of the three trophic levels of the base set (fish, zooplankton and algae) | 1000 |
| One long-term NOEC (nonobserved effect concentration, either for fish or *Daphnia*) | 100 |
| Two long-term NOECs from species representing two trophic levels | 50 |
| Long-term NOECs from at least three species (normally fish, *Daphnia*, and algae) representing three trophic levels | 10 |
| Field data or model ecosystems | case by case |

Source: From Jensen, C., Nielsen, S.N., and Halling Sørensen, B., 1998, Report (in Danish) on the environmental risk assessment of drugs, The Danish Environmental Protection Agency, Copenhagen. With permission.

assessment. Many biological processes such as colonization, reproduction, and mortality need also to be described stochastically.

Human errors are inevitably attributes of all human activities. This type of uncertainty includes incorrect measurements, data recording errors, computational errors, and so on.

The uncertainty is considered by use of an assessment factor from 10 to 1000. The choice of assessment factor depends on the quantity and quality of toxicity data (see Table 10.1). The assessment or safety factor is used in step 3 of the environmental risk assessment procedure, presented below. Relationships other than the uncertainties originating from randomness, errors, and lack of knowledge may be considered when the assessment factors are selected, for instance, cost-benefit. It implies that the assessment factors for drugs and pesticides, for example, may be given a lower value.

Lack of knowledge results in undefined uncertainty that cannot be described or quantified. It is a result of practical constraints on our ability to accurately describe, count, measure, or quantify everything that pertains to a risk estimate. Clear examples are the inability to test all toxicological responses of all species exposed to a pollutant and simplifications needed in the model used to predict the expected environmental concentration.

The most important feature distinguishing risk assessment from impact assessment is the emphasis in risk assessment on characterizing and quantifying uncertainty. It is therefore of particular interest in risk assessment to be able to analyze and estimate the analyzable uncertainty. They are natural stochasticity, parameter errors, and model errors. Statistical methods may provide direct estimates of uncertainties. They are widely used in model development (see Jørgensen, 1994).

The use of statistics to quantify uncertainty is complicated in practice by the need to consider errors in both the dependent and independent variables and to combine errors when multiple extrapolations should be made. Monte Carlo analysis is often used to overcome these difficulties (see Bartell et al., 1983).

Model errors include inappropriate selection or aggregation of variables, incorrect functional forms, and incorrect boundaries. The uncertainty associated with model errors is usually assessed by field measurements utilized for calibration and validation of the model (Jørgensen, 1994).

## 10.2 How to Perform ERA?

Risk assessment of chemicals may be divided into nine steps, which are shown in Figure 10.2.

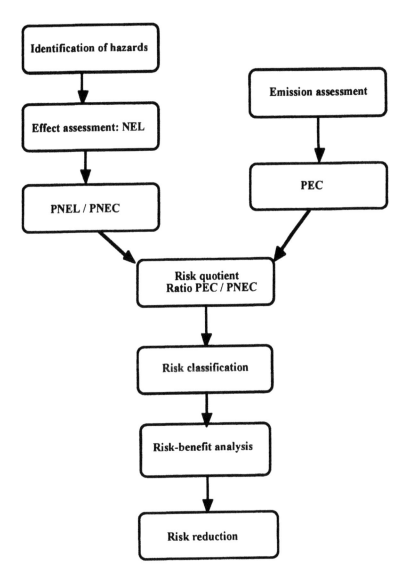

**FIGURE 10.2**
The procedure in nine steps to assess the risk of chemical compounds. Steps 1 to 3 require extensive use of ecotoxicological handbooks and ecotoxicological estimation methods to assess the toxicological properties of the considered chemical compounds, while step 5 requires a selection of a proper ecotoxicological model.

The nine steps correspond to questions the risk assessment attempts to answer to be able to quantify the risk associated with the use of a chemical. The nine steps are presented in detail below with reference to Figure 10.2.

**Step 1:** Which hazards are associated with the application of the chemical? This involves gathering data on the types of hazards — possible environmental damages and human health effects. The health effects include congenital, neurological, mutagenic, and cancerogenic effects. They may also include characterization of the behavior of the chemical within the body (interactions with organs, cells or genetic material). What is the possible environmental damage, including lethal and sublethal effects, on growth and reproduction of various populations.

As an attempt to quantify the potential danger posed by chemicals, a variety of toxicity tests has been devised. Some of the recommended tests involve experiments with subsets of natural systems, e.g., microcosms, or with entire ecosystems. The majority of testing of new chemicals for possible effects has, however, been confined to studies in the laboratory on a limited number of test species. Results from these laboratory assays provide useful information for quantification of the relative toxicity of different chemicals. They are used to forecast effects in natural systems, although their justifications have been seriously questioned (Cairns et al., 1987).

**Step 2:** What is the relation between dose and response of the type defined in step 1? It implies knowledge to NEC, $LD_x$-, $LC_y$- and $EC_z$-values where x, y, and z express a probability of harm. The answer can be found by laboratory examinations or we may use the estimations methods presented in Chapter 11. Based on these answers, a most probable level of no effect, NEL, is assessed.

Data needed for steps 1 and 2 can be obtained directly from scientific libraries but are increasingly found via on-line data searches in bibliographic and factual databases. Data gaps should be filled with estimated data (see Chapter 11).

**Step 3:** Which uncertainty (safety) factors reflect the amount of uncertainty that must be taken into account when experimental laboratory data or empirical estimation methods are extrapolated to real situations? Usually, safety factors of 10 to 10,000 are used. The choice is discussed above and will usually be in accordance with Table 10.1. If a good knowledge of the chemical is available, a safety factor of 10 may be applied. If, on the other hand, it is estimated that the available information has a high uncertainty, a safety factor of 10,000 may be recommended. Most frequently, safety factors of 50 to 100 are applied. NEL times the safety factor is called the *predicted noneffect level*, PNEL. The complexity of environmental risk assessment is often simplified by deriving *predicted no effect concentration*, PNEC, for different environmental compartment, (water, soil, air, biotas and sediment).

**Step 4:** What are the sources and quantities of emissions? The answer requires a complete knowledge of the production and use of the considered chemical compounds, including an assessment of how much of the chemical is wasted to the environment by production and use? The chemical may also be a waste product, which makes it very difficult to determine the amounts involved. For instance, the very toxic dioxins are the waste products by incineration of organic waste.

**Step 5:** What is the actual exposure concentration? The answer to this question is called the *predicted environmental concentration*, PEC. Exposure can be assessed by measuring environmental concentrations. It may also be predicted by a model, when the emissions are known. The use of models is necessary in most cases either because we are considering a new chemical, or because the assessment of environmental concentrations requires a very high number of measurements to determine the variations in concentrations in time and space. Furthermore, it provides an additional certainty to compare model results with measurements, which implies that it is always recommended to both develop a model and make at least a few measurements of concentrations in the ecosystem components, where it is expected that the highest concentration will occur. Most models will demand an input of parameters, describing the properties of the chemicals and the organisms, which also will require an extensive application of handbooks and a wide range of estimation methods. The development of an environmental, ecotoxicological model requires, therefore, an extensive knowledge of the physical–chemical–biological properties of the considered chemical compound(s). The selection of a proper model is discussed in Section 10.3.

**Step 6:** What is the ratio PEC/PNEC? This ratio is often called the *risk quotient*. It should not be considered an absolute assessment of risk but rather a relative ranking of risks. The ratio is usually found for a wide range of ecosystems, such as aquatic ecosystems, terrestrial ecosystems, and groundwater.

Steps 1 to 6 are shown in more detail with the indication of 10 steps in Figure 10.3, which is completely in accordance with Figure 10.2 and the information given above.

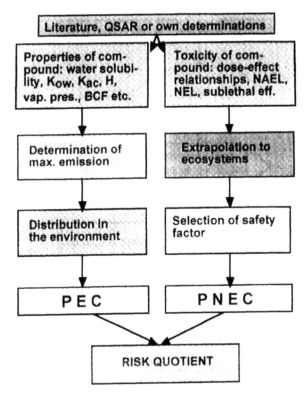

**FIGURE 10.3**
Steps 1 to 6 are shown in more detail for practical applications. The result of these steps leads naturally to assessment of the risk quotient.

**Step 7**: How will you classify the risk? The valuation of risks is made in order to decide on risk reductions (step 9). Two risk levels are defined:

1. The upper limit, i.e., the maximum permissible level (MPL)
2. The lower limit, i.e., the negligible level, NL. It may also be defined as a % of MPL, for instance 1% of MPL

The two risk limits create three zones: a black, unacceptable, high-risk zone > MPL, a grey, medium-risk level, and a white, low-risk level < NL.

Figure 10.4 shows the risk characterization and risk evaluation process for three risk assessment stages, based on the PEC/PNEC ratio (see the description of step 6). In each stage, the data set is used for exposure assessment, effect assessment, and risk characterization.

**Step 8:** What is the relation between risk and benefit? This analysis involves examinations of social–economic, political, and technical factors, which are beyond the topic of this volume. The cost–benefit analysis is difficult, because the costs and benefits are often of a different order.

**Step 9:** How can the risk be reduced to an acceptable level? The answer to this question requires a profound technical, economic, and legislative investigation. Assessment of alternatives is often an important aspect in risk reduction.

Steps 1 to 3 + 5 require a knowledge of the properties of the focal chemical compounds, which again implies an extensive literature search or a selection of the best feasible estima-

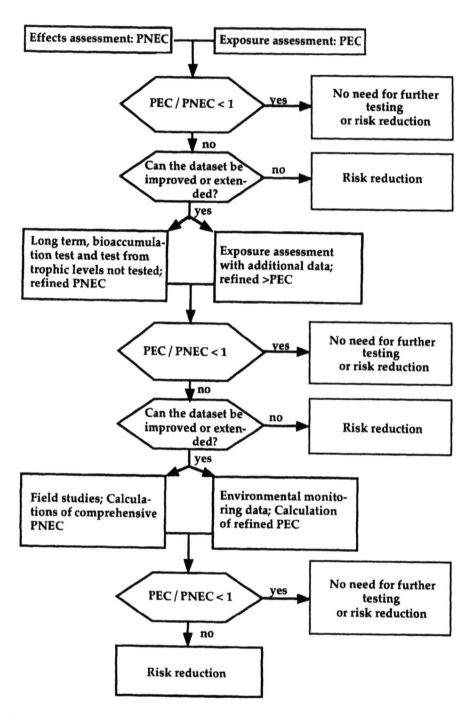

**FIGURE 10.4**
Decision tree for risk characterization and evaluation in three stages. Characterization is improved or extended.

tion procedure. In addition to Beilstein it can be recommended to have at hand the following very useful handbooks of environmental properties of chemicals and methods for estimation of these properties in case the literature values are not available:

S.E. Jørgensen, S. Nors Nielsen, and L.A. Jørgensen, *Handbook of Ecological Parameters and Ecotoxicology*, Elsevier, 1991.

P.H. Howard et al., *Handbook of Environmental Degradation Rates*, Lewis Publishers, 1991.

K. Verschueren, *Handbook of Environmental Data on Organic Chemicals*, Van Nostrand Reinhold, 1983.

P.H. Howard, Handbook of Environmental Fate and Exposure Data, Lewis Publishers. *Volume I. Large Production and Priority Pollutants*, 1989; *Volume II. Solvents*, 1990; *Volume III. Pesticides*, 1991; *Volume IV. Solvents 2*, 1993; *Volume V. Solvents 3*, 1998.

G.W.A. Milne, *CRC Handbook of Pesticides*. CRC Press, 1994.

W.J. Lyman, W.F. Reehl, and D.H. Rosenblatt, *Handbook of Chemical Property Estimation Methods. Environmental Behaviour of Organic Compounds*, American Chemical Society, 1990.

D. Mackay, W.Y. Shiu, and K.C. Ma, Illustrated Handbook of Physical–Chemical Properties and Environmental Fate for Organic Chemicals, Lewis Publishers. *Volume I. Mono-aromatic Hydrocarbons. Chloro-benzens and PCBs*, 1991; *Volume II. Polynuclear Aromatic Hydrocarbons, Polychlorinated Dioxines, and Dibenzofurans*, 1992; *Volume III. Volatile Organic Chemicals*, 1992.

S.E. Jørgensen, H. Mahler, and B. Halling Sørensen, *Handbook of Estimation Methods in Environmental Chemistry and Ecotoxicology*, Lewis Publishers, 1997.

Steps 1 to 3 sometimes denote effect assessment or effect analysis, and steps 4 to 5 exposure assessment or effect analysis. Steps 1 to 6 may be called risk identification, while environmental risk assessment, ERA, encompasses all 9 steps presented in Figure 10.2. Step 9 is particularly demanding, because several possible steps in reduction of the risk should be considered, including treatment methods, cleaner technology, and substitutes for the examined chemical.

During the last 5 or 6 years in North-America, Japan, and the EU it has been considered to treat medicinal products similar to other chemical products, as there, in principle, is no difference between a medicinal product and other chemical products. However, this initiative has only resulted in environmental risk assessment for new veterinary medicinal products. At present, technical directives for human medicinal products in the EU does not include any reference to ecotoxicology and the assessment of their potential risk. However, a detailed technical draft guideline issued in 1994 indicates that the approach applicable to veterinary medicine would also apply to human medicinal products. Presumably, ERA will be applied on all medicinal products in the near future, when sufficient experience with the veterinary medicinal products has been achieved. Veterinary medicinal products are released in larger amounts to the environment, because manure, in spite of its possible content, is utilized as fertilizer on agricultural fields.

It is also possible to perform an environmental risk assessment with the focus on the human population. The 10 steps corresponding to Figure 10.3 are shown in Figure 10.5, which is not significantly different from Figure 10.3. The principles for the two types of environmental risk assessment are the same. Figure 10.5 uses nonadverse effect level (NAEL), nonobserved adverse effect level (NOAEL) to replace predicted noneffect concentration, and predicted environmental concentration is replaced by tolerable daily intake (TDI).

This type of environmental risk assessment has particular interest for veterinary medicine because some animal medicines may contaminate food products for human consumption. For instance, the use of antibiotics in pig feed has attracted a lot of attention, because they can be found as residue in pork or can contaminate the environment through the application of manure as natural fertilizer.

Selection of a proper ecotoxicological model is a first step in development of an environmental exposure model, as required in step 5. This will be discussed in more detail in the next section.

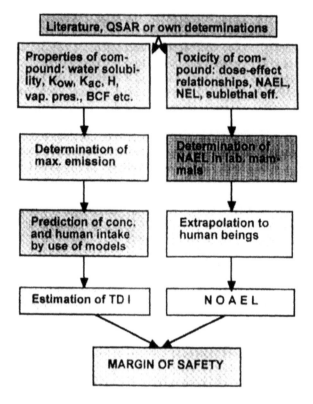

**FIGURE 10.5**
Environmental risk assessment for human exposure. It leads to a margin of safety which corresponds to the risk quotient in Figures 10.3 and 10.4.

## 10.3   Selection of Ecotoxicological Models

Toxic substance models attempt to model the fate and effect of toxic substances in ecosystems. Toxic substance models differ from other ecological models in that:

1. The need for parameters to cover all possible toxic substance models is great, and general estimation methods are therefore used widely.
2. The safety margin should be high, for instance, expressed as the ratio between the actual concentration and the concentration that gives undesired effects.
3. They require possible inclusion of an effect component, which relates the output concentration to its effect. It is easy to include an effect component in the model; it is, however, often a problem to find a well-examined relationship to base it on.
4. They need to be simple models due to points 1 and 2, and our knowledge of process details, parameters, sublethal effects, and antagonistic and synergistic effects is limited.

It may be an advantage to clarify several questions before developing a toxic substance model:

1. Obtain the best possible knowledge about the processes of the considered toxic substances in the ecosystem. As far as possible, knowledge about the quantitative role of the processes should be obtained.

2. Attempt to get parameters from the literature and/or from experiments (*in situ* or in the laboratory)

3. Estimate all parameters by use of suitable estimation methods and follow the recommendations given in Chapter 11.

4. Estimate which processes and state variables it would be feasible and relevant to include in the model. If there is the slightest doubt, include too many processes and state variables rather than too few.

5. Use a sensitivity analysis to evaluate the significance of the individual processes and state variables. This often may lead to further simplifications.

Ecotoxicological models can be divided into six classes. The classification presented here is based on differences in the modeling structure. The decision of which model class to apply is based upon the ecotoxicological problem that the model aims to solve. The definitions of the model classes are given below.

1. **Food chain or food web dynamic models.** This class of models considers the flow of toxic substances through the food chain or food web. Such models will be relatively complex and contain many state variables. The models will also contain many parameters, which often have to be estimated by one of the methods presented in this volume. This type of model will typically be used when many organisms are affected by the toxic substance, or the entire structure of the ecosystem is threatened by the presence of a toxic substance. Because of the complexity of these models, they have not been widely used. They are similar to the more complex eutrophication models that consider the flow of nutrients through the food chain or even through the food web.

2. **Static models of the mass flows of toxic substances.** If the seasonal changes are minor, or of minor importance, a static model of the mass flows will often be sufficient to describe the situation and even to show the expected changes if the input of toxic substances is reduced or increased. It will often, but not necessarily, contain more trophic levels, but the modeler is frequently concerned with the flow of the toxic substance through the food chain. If there are some seasonal changes, this type, which usually is simpler than type one, can still be an advantage to use, for instance, if the modeler is concerned with the worst case and not with the changes.

3. **A dynamic model of a toxic substance in a trophic level.** It is often only the toxic substance concentration in one trophic level is of concern. This includes the zero trophic level, which is understood as the medium — soil, water, or air. The simplifications behind the use of this type of model are often feasible when the problem is well defined, including which component is most sensitive to toxic matter, and which processes are most important for concentration changes.

4. **Ecotoxicological models in population dynamics.** Population models are biodemographic models and therefore have numbers of individuals or species as state variables. The simple population models consider only one population. The growth of the population is a result of the difference between natality and mortality:

$$dN/dt = B*N - M*N = r*N, \qquad (10.1)$$

where N is the number of individuals, B is the natality, i.e., the number of new individuals per unit of time and per unit of population, M is the mortality, i.e., the number of organisms that die per unit of time and per unit of population; and r is the increase in the number of organisms per unit of time and per unit of population, and is equal to B − M. B, N, and r are not necessarily constants as in the exponential growth equation, but are dependent on N, the carrying capacity, and other factors. The concentration of toxic substance in the environment or in the organisms may also influence the natality and the mortality, and if the relation between a toxic substance concentration and these population dynamic parameters is included in the model, it becomes an ecotoxicological model of population dynamics.

Population dynamic models may include two or more trophic levels, and ecotoxicological models will include the influence of the toxic substance concentration on natality, mortality, and interactions between these populations. In other words, an ecotoxicological model of population dynamics is a general model of population dynamics with the inclusion of relations between toxic substance concentrations and some important model parameters.

5. **Ecotoxicological models with effect components.** Although class 4 models already include relations between concentrations of toxic substances and their effects, these are limited to population dynamic parameters. In comparison, class 5 models include more comprehensive relations between toxic substance concentrations and effects. These models may include not only lethal and/or sublethal effects but also effects on biochemical reactions or on the enzyme system.

In many problems it may be necessary to go into more detail on the effect to answer the following relevant questions:

1. Does the toxic substance accumulate in the organism?

2. What will be the long-term concentration in the organism when uptake rate, excretion rate, and biochemical decomposition rate are considered?

3. What is the chronic effect of this concentration?

4. Does the toxic substance accumulate in one or more organs?

5. What is the transfer between various parts of the organism?

6. Will decomposition products eventually cause additional effects?

A detailed answer to these questions may require a model of the processes that take place in the organism, and a translation of concentrations in various parts of the organism into effects. This implies, of course, that the intake = (uptake by the organism)*(efficiency of uptake) is known. Intake may either be from water or air, which also may be expressed by concentration factors, which are the ratios between the concentration in the organism and in the air or water.

However, if all the above-mentioned processes were to be taken into consideration for just a few organisms, the model would easily become too complex, contain too many parameters to calibrate, and require more detailed knowledge than it is possible to provide. Often we do not even have all the relations needed for a detailed model, as toxicology and ecotoxicology are still in their infancy. Therefore, most models in this class will not consider too many details of the partition of the toxic substances in organisms and their corresponding effects, but rather be limited to the simple accumulation in the organisms and their effects. Usually, accumulation is rather easy to model and the following simple equation is often sufficiently accurate:

$$d\,C/d\,t = (ef*Cf*F + em*Cm*V)/W - Ex*C = (INT)/W - Ex*C \qquad (10.2)$$

where C is the concentration of the toxic substance in the organism; ef and em are the efficiencies for the uptake from the food and medium, respectively (water or air); Cf and Cm are the concentration of the toxic substance in the food and medium, respectively; F is the amount of food uptake per day; V is the volume of water or air taken up per day; W is the body weight either as dry or wet matter; and Ex is the excretion coefficient (1/day). As can be seen from the equation, INT covers the total intake of toxic substance per day.

This equation has a numerical solution:

$$C/C(max) = (INT*(1 - \exp(Ex*t)))/(W*Ex) \qquad (10.3)$$

where C(max) is the steady state value of C:

$$C(max) = INT/(W*Ex) \qquad (10.4)$$

Synergistic and antagonistic effects have not been touched on so far. They are rarely considered in this type of model for the simple reason that we do not know much about these effects. If we have to model combined effects of two or more toxic substances, we can only assume additive effects, unless we can provide empirical relationships for the combined effect.

6. **Fate models with or without a risk assessment component.** The last type of ecotoxicological model focuses on the fate of the toxic substances — where in the ecosystem will the toxic substance be found? In what concentration?

**Fugacity models** are special types of fate model. They attempt to answer the following questions: In which of six compartments, corresponding to the spheres, can we expect the greatest problem for chemicals emitted to the environment? Which concentration will be expected in each compartment? What are the implications of this concentration? The fugacity models are to a large extent based on physical–chemical parameters and the use of these parameters to estimate the other required parameters. The applicability of the fugacity model is therefore very dependent on the use of the estimation methods presented in this volume.

Information on development of fate models, particularly fugacity models, can be found in D. Mackay, 1991, *Multimedia Environmental Models: The Fugacity Approach*. Mackay distinguishes four levels of fate (fugacity, multimedia) models. This particular approach is presented in Mackay and Paterson (1982).

The first level calculates the equilibrium distribution of a chemical between phases. It assumes that each compartment is well mixed and there is no reaction or advection into or out of the considered system.

The second level considers equilibrium but also includes reaction and advection. Reaction comprises photolysis, hydrolysis, biodegradation, oxidation, and so on. All the processes are assumed to be first-order reactions.

The third level is devoted to a steady-state nonequilibrium situation, which implies that the fugacities are different in each phase.

Level four involves a dynamic version of level three, where emissions and thus concentrations vary with time.

Levels one or two are usually sufficient, but if the ecotoxicological management problem requires prediction of the time needed for a substance to accumulate to a certain concentration in a phase after emission has started or the length of time for the system to recover after the emission has ceased, then a level four model should be applied.

SETAC, in 1995, published a review of multimedia fate models in volume named *The Multi-Media Fate Model: A Vital Tool for Predicting the Fate of Chemicals.* The review includes a comparison of four available software programs based on multimedia fate models: HAZCHEM, ChemCAN, SimpleBOX, and CalTOX.

The review concludes that all four software packages are recommended for the development of multimedia predictions, although a proper validation in the sense applied in ecological modeling has not been carried out.

Multimedia fate models are useful in risk assessment of chemicals (compare with the procedure presented in Figure 10.2). If the risk assessment focuses on an ecosystem or a region, comprising a few ecosystems or types of ecosystems, class one through five models should be preferred, as they in most cases will yield more accurate predictions. Which one of these five classes to apply depends on the problem and our knowledge about the system and the processes in general. If, on the other hand, the risk assessment is concerned with an entire country or state, the multimedia fate model seems to be the only alternative. Also in chemical ranking and scoring, where two or more chemicals are compared, multimedia fate models are recommended as the proper tool. Application of level two, or perhaps level three, fugacity models is recommended for comparison of the environmental fate of alternative chemicals.

The presentation of the six classes of models above clearly shows the advantages and limitations of ecotoxicological models. The simplifications used in classes two, three, and six (at least without risk assessment components and levels 1 to 2 of Mackay fugacity models) often offer great advantages. They are sufficiently accurate to give a good overview of the concentrations of toxic substances in the environment, due to the application of high safety factors. The application of the estimation methods renders it feasible to construct such models, even our knowledge of the parameters is limited. The estimation methods obviously have a high uncertainty, but the great safety factor helps in accepting this uncertainty. On the other hand, our knowledge about the effects of toxic substances is very limited — particularly at the organism and organ level. It must not be expected, therefore, that models with effect components give more than a rough picture of what is known today in this area.

Another classification of ecotoxicological models should also be mentioned, because it illustrates very clearly the thoughts behind modeling in environmental chemistry. The classification uses three classes of models, as also presented in Chapter 2:

A. Fate models, included fugacity models
B. Ecosystem specific models of toxic substances
C. Models of toxic substances in a typical or average ecosystem which is used as a general representation of all ecosystems of a given type

**Class A models** are used to get a rough estimation of where a toxic substance will be found in the environment and at approximately that concentration. This class of ecotoxicological models is useful for comparing various alternative chemicals. It may be used to answer the following question: Should we prefer chemical X or Y from an environmental point of view? This type of model has been treated comprehensively in D. Mackay's book *Multimedia Environmental Models.*

**Class B models** are used when we have to decide on abatement of a specific pollution of a toxic substance, for instance when an organic substance is transported from a chemical plant by wastewater to the environment, perhaps after the wastewater has passed a treatment plant. In this case it is necessary to include the characteristic features of the ecosystems receiving the toxic substance. This type of ecotoxicological model is obviously similar to ecological models, because both must include the same characteristic ecological features. This class of model is generally more accurate than class A models, but it is also more specific and can hardly be applied to ecosystems other than that for which it was developed. This class of models is presented in details in Jørgensen (1994).

**Class C models** have recently found increasing application because many chemicals are discharged to specific types of ecosystems. If we want, for instance, to select a pesticide that is harmless to earthworms, we have to "test" our spectrum of pesticides on a general agricultural ecosystem model with a food chain that includes earthworms. We don't have a specific agricultural system in mind, but a very general agricultural ecosystem with average features characteristic for such systems. In this case, we will develop a model with the characteristic features of an average agricultural system. This class of models will be less accurate than class B models but will generally give more accurate and more specifically applicable results than class A models. On the other hand, the results cannot be applied with the same generality as for class A models.

It is recommended to use all three classes of models to assess the environmental risk of application of veterinary medicinal products. They emphasize the need for calculations of PEC on both a local and a regional spatial scale, preferably from monitoring data.

It is generally very difficult to find data for calibration and validation of class A models, while data for calibration and validation of class B models always should be provided to assure an acceptable accuracy of the model results. It may be possible to get data for calibration and validation of class C models, but the calibration and validation cannot be carried out with the same accuracy as for class B models, because the data will be valid only for specific ecosystems.

It is always valuable in the development of a new ecotoxicological model to build on the experience already gained in this area. A summary of the available ecotoxicological models can be found in S.E. Jørgensen, B. Halling Sørensen, and S.N. Nielsen, *Handbook of Environmental and Ecological Modeling* (1995). The handbook gives a short description of more than 400 models, including 71 ecotoxicological models: 16 models of pesticides, 28 models of other organic compounds, and 27 models of inorganic compounds (heavy metals and radionuclides). The 71 models encompass all six classes of models mentioned above.

For each model is given: model identification, model type, model purpose, a short description of state variables and forcing functions, model application, model software and hardware demands, model availability, model documentation and major references, and other information relevant for other modelers. It can often save a lot of time in the development of models to learn from others' experience. It is of particular importance to touch on the following modeling features:

- What is a proper model complexity?
- Which state variables are important to include in the model?
- Which process descriptions are working properly?

Answers to these questions can be found in the above-mentioned handbook, which is published with the aim of facilitating the model development for modelers by applying experience from more than 400 different models.

## 10.4   Discussion of Specific Examples of Risk Assessment

Environmental risk assessment follows the nine steps shown in Figure 10.2, but each of the steps can be performed in more or less detail (see also Figure 10.3 and 10.5). A wide spectrum of possibilities for each of the nine steps is offered and has been tested in practice. Some of the most useful approaches are illustrated below.

### 10.4.1   Chlorinated Aliphatic Hydrocarbons

The first example to illustrate the application of risk assessment of chemicals is associated with the discharge of a mixture of chlorinated aliphatic hydrocarbons, either straight chained or branched, with chain lengths that range between 10 and 30 carbon atoms. It is based on a similar case study presented by Bartell et al. (1992). Commercial use includes additives for lubricating oils, plasticizers, traffic paints, and additives for flame retardants. Approximately 15,000 metric tons enter the environment annually, mainly to the hydrosphere, as these chemical compounds are discharged mainly into wastewater.

The steps in the example of environmental risk assessment are summarized in Figure 10.6. They are partly proposed by Bartell et al. (1992) and represent a more elaborate edition of risk assessment, as more modeling steps than usual are used.

**Step 1.** It is recommended to develop a model with the scope to predict the final concentration in surface water. It should consider biodegradation in a typical wastewater treatment plant and the adsorption equilibrium between sludge and water (step 1A). The further fate of the chemicals in freshwater ecosystems should be covered by a model which includes uptake by aquatic organisms and the adsorption equilibrium between water and sediment (step 1B). It is not the aim here to give the details of the model's construction. Those who are interested in the problems associated with selection of the right model and its development for practical use in environmental management and environmental risk assessment are referred to Jørgensen (1994).

Model results **(step 2)** and analyses **(step 3)** indicate that concentrations in surface water associated with industrial inputs may range between 0.5 and 6 ppb, which is about 10 times the background concentration (see Jørgensen et al., 1991). The concentration in sediment is modeled and reported as much as 1000 to 3000 times greater than dissolved concentrations. Due to their lipophilic character, bioaccumulation in aquatic organisms is considered a possible environmental problem.

**Step 4.** It is useful to try to obtain knowledge about all properties of the range of chemicals covered by the label "chlorinated aliphatic hydrocarbons" to get as complete an image as possible of the environmental consequences that can be expected from use of these chemicals (see also Section 10.2).

The $EC_{50}$-concentration for phytoplankton growth is known to be 31.6 µg/l (96h) (see Jørgensen et al., 1991), $LC_{50}$ for zooplankton is 46 µg/l, and for fish of different species an $LC_{50}$ of 100 to 300 mg/l may be applicable (fathead minnow 100 mg/l and bluegill sunfish 300 mg/l). It is recommended that an exposure model be used that will relate concentrations in surface water to their effects on the aquatic organisms. This is a useful method for filling the gap in our knowledge about the detailed relation between concentrations and effects.

**Step 5.** If we only have one point on the response exposure curve (see Figure 10.7), we are forced to use a linear approximation to find other exposure/response relations. We use this linear relation to find the effect for any concentration.

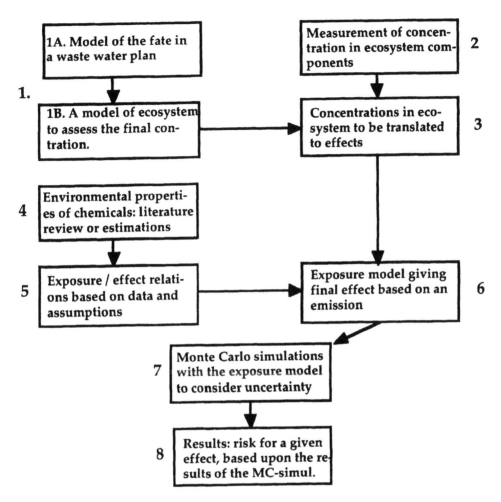

**FIGURE 10.6**
The 8 steps in the specific example, chlorinated aliphatic hydrocarbons in aquatic ecosystems, of environmental risk assessment. Models and/or analyses are presumed to be applied to find the actual range of concentrations. The environmental properties are translated into an effect/exposure relation, where in most cases it is necessary to assume the relationship (linear, exponential etc.) to be able to interpolate or extrapolate to other exposures. The uncertainty of the parameters in the exposure model is used to perform Monte Carlo simulations, which can give the relation between the probability of a risk for an effect.

**Step 6.** The growth of phytoplankton in the exposure model may be expressed by an equation:

$$dB/dt = f \text{ (light, nutrients, respiration, temperature, sinking, mortality, grazing)} \quad (10.5)$$

where B is the biomass. The usual expressions for growth as a function of light and nutrients, Michaelis–Menten equations, are used. The sinking and mortality are first-order reactions and grazing is a Michaelis–Menten expression. The concentration of the grazers is regulated by predation of planktivorous fish using a general Michaelis–Menten expression. The influence of the temperature is an exponential function as generally applied.

The influence of the toxic substance by this method presumes that the various parameters are equally influenced by a factor, x, to obtain the expected effect. If we, for instance, want to find the parameters valid for 31.6 µg/l chlorinated aliphatic hydrocarbons, we should find the x value which corresponds to half the increase in biomass found by integration of Equation 10.5 over a period of 96 h (three days). The growth rate is x times the

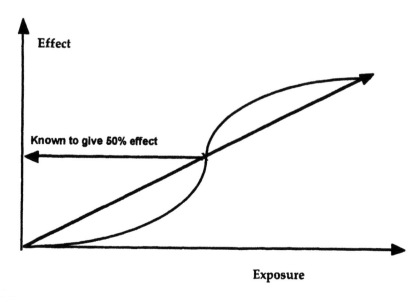

**FIGURE 10.7**

Typical relationship between effect and exposure. If only one set of related exposure to effect is known, it is necessary to presume a linear relationship as shown. It cannot be denied, of course, that the relation could be different, as for instance the logistic-like curve shown in the figure.

previous growth rate; the half saturation constant for the influence of nitrogen and phosphorus on growth in the Michaelis–Menten expression is the previous value divided by x; the half saturation constant for light is x times the previous value; the respiration coefficient, sinking rate, and mortality rate are the previous values divided by x; and the grazing rate is also x times less than the previous value. x is found through iterations by trial and error. For a concentration of 31.6 µg/l, an x value giving a growth rate which is 75% of normal can be found, and so on.

**Step 7.** All the parameters in the exposure model have, however, a standard deviation rooted in the uncertainty of our knowledge of the value of the model parameters, including the parameters translating exposure into effect. The exposure model and the translation concentration → effect are run by Monte Carlo simulations based on the estimated uncertainties, perhaps 1000 times for each of the actual concentrations. If a certain effect, say 25% reduction of zooplankton population, is obtained by the simulations in 10% of the cases, it is concluded that the risk of 25% reduction in the zooplankton concentration is 10%.

**Step 8.** The results of this approach to model the effect of chlorinated aliphatic hydrocarbons on phytoplankton, zooplankton, and fish are summarized in Table 10.2 (more

**TABLE 10.2**

Results of the Environmental Risk Assessment for Chlorinated Aliphatic Hydrocarbons by Use of an Exposure Model. The Probability for the Indicated Risk Is Shown.

| Concentration µg/l | Algal Increase | | Zooplankton Decrease | | Fish Decrease | |
|---|---|---|---|---|---|---|
| | 200% | 400% | 25% | 50% | 25% | 50% |
| 0.1 | 0.33 | 0.14 | 0.0064 | ≈ 0 | 0.01 | ≈0 |
| 1 | 0.70 | 0.45 | 0.05 | ≈ 0 | 0.40 | 0.043 |
| 10 | 0.05 | 0.03 | 0.14 | 0.088 | 0.95 | 0.89 |
| 50 | 0.006 | 0.006 | 0.075 | 0.044 | 0.94 | 0.90 |
| 100 | 0.008 | 0.002 | 0.091 | 0.065 | 0.94 | 0.87 |

detailed results are presented in Bartell et al., 1992, where different scenarios are presented). Notice that the algal increase is considered to be caused mainly by the indirect effect of the decrease in zooplankton. An exposure model is applied because it is often impossible to predict the expected effects in an ecosystem due to the complex relations. It is therefore strongly recommended to apply exposure models whenever complex relations on the ecosystem levels are considered.

### 10.4.2 Risk Assessment of Benzene

This example is concerned with the risk assessment on human beings. The properties of benzene can easily be found in handbooks. Its most important properties are as follows:

1. Water solubility 1.8 g/l
2. Log $K_{ow}$ = 2.3
3. Log BCF = 1 to 3
4. Log $K_{oc}$ = 1.9
5. Vapor pressure at room temperature: 12 700 Pa

Based on these properties, a fate model according to Mackay (1991) can be developed. It shows that 99% of benzene released to the environment will be found in the atmosphere.

Several effects on humans are known for benzene: reduced production of hemoglobin, narcosis, toxic effect on the central nerve system, and it is cancerogenic. The relation between exposure and effects is shown in Figure 10.8. Experiments on rats have shown that the no-observed-effect concentration in air is approximately 100 µg/l for long-term exposure (7 hours per day, 5 days per week). It is converted to constant exposure by multiplication by 7 × 5 and division of 24 by 7, and we get approximately 20 µg/l. If a safety factor of 500 is used, PNEC becomes 0.04 µg/l, corresponding to 0.014 ppm.

Howard (1990) provides information about a relationship between the relative increase in mortality RIM (%) due to blood cancer and the exposure to benzene:

$$RIM = bxd \tag{10.6}$$

where b is a constant and d is the exposure expressed as ppm times the number of years. A b value of 1 is suggested.

Measurements of benzene concentrations show that in streets and roads with heavy traffic concentrations in the range of 1 to 16 ppb are found. It yields a ratio of PEC/PNEC close to or slightly greater than 1. RIM for a person exposed to 16 ppb = 0.016 ppm in 70 years is 1 × 70 × 0.016 = 1.2%, which is unacceptable. It can be concluded that a reduction of the risk is required.

### 10.4.3 Risk Assessment of Bisphenol-A (BPA)

This example is concerned with the effects on various organisms. The most important properties are:

1. Boiling point: 250°C
2. Water solubility at room temperature: 120 mg/l
3. Henry's constant 0.00001 Pa m³/mol

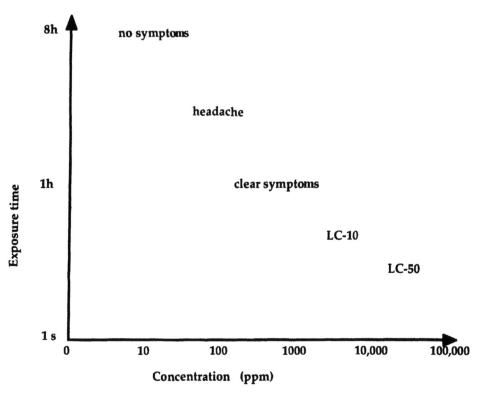

**FIGURE 10.8**

The relation between effect and a combination of concentration and exposure time is shown. Notice that exposure is considered a combination of concentration and time.

4. $\log K_{ow}$ = 2.2 to 3.4
5. $\log K_{oc}$ = 2.5 to 3.1
6. BCF = 5 to 70 (fish)

Effect/exposure data:

1. Effect on growth rate of algae of a concentration of 1 to 2 mg/l
2. $LC_{50}$ for freshwater fish species ≥ 4.4 mg/l
3. Estrogen effects may be observed at 2.3 μg/l (Andersen, 1998)

A fate model shows that the major part of BPA emissions to the environment will be present in the hydrosphere. A type C model of a freshwater system (see also steps 1A and 1B in Figure 10.4) which receives industrial discharge of BPA via a wastewater treatment plant results in a concentration of about 0.5 to 2 μg/l in typical rivers and small lakes. Measurements from freshwater systems with high risk indicate concentrations in the range of 1 to 10 μg/l.

A PNEC of 1 μg/l has been suggested, based upon an NOEL of 1 mg/l and a safety factor of 1000. This value doesn't consider the estrogen effect, which would give a much lower PNEC effect even using a safety factor of only 100. A ratio of PEC/PNEC of 1 to 10 is found, when PNEC = 1 μg/l is applied. As BPA is used in considerable amounts, a reduction of the emission to the environment is urgently needed. This can be concluded on the basis of this relatively simple environmental risk assessment.

# References

Andersen, H.R., 1998, Master's thesis in Danish at DFH, Environmental Chemistry: *Examination of Endocrine Disruptors.*

Bartell, S.M., Gardner, R.H., and O'Neill, R.V., 1992, *Ecological Risk Estimation*, Lewis Publishers, Boca Raton.

Cairns J., Jr., Dickson, K.L., and Maki, A.W., 1987, *Estimating Hazards of Chemicals to Aquatic Life*, STP. 675, American Society for Testing and Materials, Philadelphia.

Calow, P., 1998, Ecological risk assessment: Risk for what? How do we decide? *Ecotoxicology and Environmental Safety*, 40: 15-18.

Chichilnisky, G., Neal, G.M., and Vercelli, A., 1997, *Sustainability: Dynamics and Uncertainty*, Kluwer Academic Publishers, Dordrecht.

Howard, P.H., et al., 1991, *Handbook of Environmental Degradation Rates*, Lewis Publishers, New York.

Howard, P.H., *Handbook of Environmental Fate and Exposure Data*, Lewis Publishers, New York, *Volume I, Large Production and Priority Pollutants*, 1989; *Volume II, Solvents*, 1990; *Volume III, Pesticides*, 1991; *Volume IV, Solvents 2*, 1993; *Volume V, Solvents 3*, 1998.

Jensen, C., Nielsen, S.N., and Halling-Sørensen, B., 1998, Report (in Danish) on the environmental risk assessment of drugs, The Danish Environmental Protection Agency, Copenhagen.

Jørgensen, S.E., 1994, *Fundamentals of Ecological Modelling*, 2nd ed., *Developments in Environmental Modelling*, 19, Elsevier, Amsterdam.

Jørgensen, S.E., Nors Nielsen, S., and Jørgensen, L.A., 1991, *Handbook of Ecological Parameters and Ecotoxicology*, Elsevier, Amsterdam.

Jørgensen, S.E., Halling Sørensen, B., and Nielsen, S.N., 1995, *Handbook of Environmental and Ecological Modeling*, CRC/Lewis Publishers, Boca Raton.

Jørgensen, S.E., Mahler, H., and Halling Sørensen, B., 1997, *Handbook of Estimation Methods in Environmental Chemistry and Ecotoxicology*, Lewis Publishers, Boca Raton.

Lyman, W.J., Reehl, W.F., and Rosenblatt, D.H., 1990, *Handbook of Chemical Property Estimation Methods. Environmental Behaviour of Organic Compounds*, American Chemical Society, Washington, D.C.

Mackay, D., 1991, *Multimedia Environmental Models: The Fugacity Approach*, Lewis Publishers, Boca Raton.

Mackay, D. and Paterson, S., 1982, Fugacity revisited, *Environ. Sci. Technol.*, 16, 654A.

Mackay, D., Shiu, W.Y., and Ma, K.C., *Illustrated Handbook of Physical–Chemical Properties and Environmental Fate for Organic Chemicals*, Lewis Publishers, Boca Raton. *Volume I. Mono-aromatic Hydrocarbons. Chloro-benzens and PCBs*, 1991; *Volume II. Polynuclear Aromatic Hydrocarbons, Polychlorinated Dioxines, and Dibenzofurans*, 1992; *Volume III. Volatile Organic Chemicals*, 1992.

Milne, G.W.A., 1994, *CRC Handbook of Pesticides.* CRC Press, Boca Raton.

SETAC, 1995, *The Multi-Media Fate Model: A Vital Tool for Predicting the Fate of Chemicals*, Pensacola, FL.

Suter, G.W., 1993, *Ecological Risk Assessment*, Lewis Publishers, Chelsea, MI.

Verschueren, K., 1983, *Handbook of Environmental Data on Organic Chemicals*, Van Nostrand Reinhold, New York.

# 11

---

## *Application of QSAR in Environmental Management*

S.E. Jørgensen and Bent Halling Sørensen

### CONTENTS

---

## 11.1   The Need for Estimation Methods

Slightly more than 100,000 chemicals are produced in such an amount that they are threatening or may threaten the environment. They cover a wide range of applications: household chemicals, detergents, cosmetics, medicines, dye stuffs, pesticides, intermediate chemicals, auxiliary chemicals in other industries, additives to a wide range of products, chemicals for water treatment and so on. They are (almost) indispensable in modern society and cover some more or less essential needs in the industrialized world, which has increased the production of chemicals about 40-fold during the last four decades. A minor or major fraction of these chemicals is inevitably reaching the environment during their production, during their transportation from the industries to the end user, or in their application. In addition, the production or the use of chemicals may cause more or less unforeseen waste or by-products. An example is chloro-compounds in the use of chlorine for disinfection. We would like to have the benefits of using these chemicals but do not accept the harm they may cause. This conflict raises several urgent questions for which we have to find answers:

1. How do these chemicals interfere with the environment? What reactions will they cause with the abiological and biological components in the environment?

2. What is the fate of these chemicals? What will be the concentrations of these chemicals and in which parts of the environment will they be found?

3. What harmful effects will these concentrations cause? Can we alleviate these harmful effects? How?

4. How much should we reduce the emissions of the chemicals to be certain of reaching a nonharmful level?

1-56670-337-9/00/$0 00+$.50
© 2000 by CRC Press LLC

5. How can we manage the production, transportation, application, and the waste products of the chemicals to obtain a nonharmful level?

6. Should the chemicals be banned if we cannot give a satisfactory answer to questions 4 and 5? Can we find a substitute for the harmful chemicals?

7. Could we assess the risk of using these chemicals? (See also Chapter 10.)

We cannot answer these crucial questions without knowing the properties of the chemicals. The OECD (Organization for Economic Cooperation and Development) has made a review of the properties that we should know for all chemicals (see Jørgensen and Johnsen, 1989; Jørgensen, 1991). We need to know the boiling point and melting point to know in which form (solid, liquid, or gas) the chemical will be found in the environment. We must know the distribution of the chemicals in the five spheres: the hydrosphere, the atmosphere, the lithosphere, the biosphere, and the technosphere. We need knowledge of the solubility in water, the partition coefficient water/lipids, Henry's constant, the vapor pressure, the rate of degradation by hydrolysis, photolysis, chemical oxidation, microbiological processes, and the adsorption equilibrium between water and soil — all as functions of the temperature. We need to reveal the interactions between the living organisms and the chemicals, which implies that we should know the biological concentration factor (BCF), the magnification through the food chain, the uptake rate and the excretion rate by the organisms, and where in the organisms the chemicals will be concentrated, not only for one organism but for a wide range. We must also know the effects on a wide range of different organisms. It means that we should be able to come up with the $LC_{50}$ and $LD_{50}$-values, the MAC and NEC-values (for the abbreviations and the definitions used, see Glossary), the relationship between the various possible sublethal effects and the concentrations, the influence of the chemical on fecundity, and its carcinogenic and teratogenic properties. We should also know its effect at the ecosystem level. How do the chemicals effect the populations and their development and interactions, i.e., the entire network of the ecosystem?

The list of properties needed to give a proper answer to the six questions mentioned above could easily be made longer (see the list recommended by OECD, and Jørgensen and Johnsen, 1989). To provide all the properties corresponding to the list given here is a huge task. More than 10 basic properties should be known for all 100,000 chemicals, which would require 1,000,000 pieces of information. In addition, we need to know at least 10 properties to describe the interactions between 100,000 chemicals. Let us say that we modestly use 10,000 organisms to represent the approximately 10 million species on Earth. This totals 1,000,000 + 100,000*10,000*10 = on the order of $10^{10}$ properties to be quantified! Today we have determined less than 1% of these properties by measurements, and with the present rate of generating new data we can be certain that during the coming century, we shall not be able to reach 10% even with an accelerated rate of ecotoxicological measurements.

With all the properties in hand, we will have to describe the processes and components involved in all the emissions of the 100,000 chemical compounds (Jørgensen, 1990, 1991). The determination of the ecotoxicological consequences of the chemicals in use is most probably another example of what Wolfram (1984a, b) called an irreducible system. The system is too complex to be measured and computed in all details. It is also consistent with the quantum mechanic idea, that it is *not* possible to give a deterministic–analytical description of the world (see Chapters 2 and 3 and Jørgensen, 1994, 1995, 1997, where this issue is discussed in an ecological context). Consequently, we have to give a holistic description of the fate of chemicals in the environment, which means that we should develop computer models and make simulations on the behavior of the considered chemicals in nature. Thereby, we shall not be able to give all the details of all the processes in nature and all the interactions with all living organisms, but we shall be able to give an approximate picture (a map) of the pathways that the chemicals follow and indicate where they probably do most harm. Many ecotoxicological models have been developed (Jørgensen et al., 1995) as answers to this challenge.

They require, however, among other inputs, information about the properties of the chemicals and their interactions with living organisms. It is not necessary to know the properties with the very high accuracy that can be provided by measurements in a laboratory, but it would be beneficial to know the properties with enough accuracy to utilize the ecotoxicological models for management and for risk assessments. Therefore, estimation methods have been developed as an urgently needed alternative to measurements. They are, to a high extent, based on the structure of the chemical compounds, the so-called QSAR and SAR methods, but it may also be possible to use allometric principles to transfer rates of interaction processes and concentration factors between a chemical and one or a few organisms to a wider range of organisms. This chapter focuses on these methods and attempts to give a brief overview on how these methods can be applied and what approximate accuracy they can offer. A more detailed overview of the methods can be found in Jørgensen et al., 1997.

It may be interesting here to discuss the obvious question: Why is it sufficient to estimate a property of a chemical in ecotoxicological context with 20%, sometimes with 50% or higher, uncertainty? Ecotoxicological models usually give an uncertainty of the same order of magnitude, which means that the indicated uncertainty may be sufficient from the modeling viewpoint, but can model results with such an uncertainty be used at all? The answer in most (many) cases is "yes," because we want in most modeling cases to assure that we are (very) far from a harmful or very harmful level. We use (see also Chapter 10 on Risk Assessment) a safety factor of 100 to 1000 in many cases. When we are concerned with very harmful effects (e.g., complete collapse of an ecosystem or a health risk for a large human population), we will inevitably select a very high safety factor. In addition, our lack of knowledge about synergistic effects and the presence of many compounds in the environment at the same time force us to apply a very high safety factor. In such context we will usually go for a concentration in the environment which is magnitudes lower than a slightly harmful effect or considerably lower than the NEC. A parallel is civil engineers constructing a bridge. They make very sophisticated calculations (develop models) that account for wind, snow, temperature changes, and so on and afterwards they multiply the results by a safety factor of 2 to 3 to ensure that the bridge does not collapse. They use safety factors because the consequences of a bridge collapse are unacceptable. A collapse of an ecosystem or a health risk to a large human population is also completely unacceptable. So, we should use safety factors in ecotoxicological modeling to account for uncertainty. Due to the complexity of the system, the simultaneous presence of many compounds and our present knowledge, or rather lack of knowledge, we should use 10 to 100, or even sometimes 1000, as a safety factor. If we use safety factors that are too high, the risk is only that the environment will be less contaminated, perhaps at a higher cost. In addition, there are no alternatives to the use of safety factors. We can increase our ecotoxicological knowledge step-by-step, but it will take decades before it may be reflected in lower safety factors. A measuring program of all processes and components is an impossibility due to the high complexity of ecosystems. Of course, this does not imply that we should not use the information of measured properties available today. Measured data will almost always be more accurate than estimated data. Furthermore, the use of measured data within the network of estimation methods will improve the accuracy of estimation methods. Several handbooks on ecotoxicological parameters are available. References to the most important ones have already been given in Chapter 10, Section 10.2.

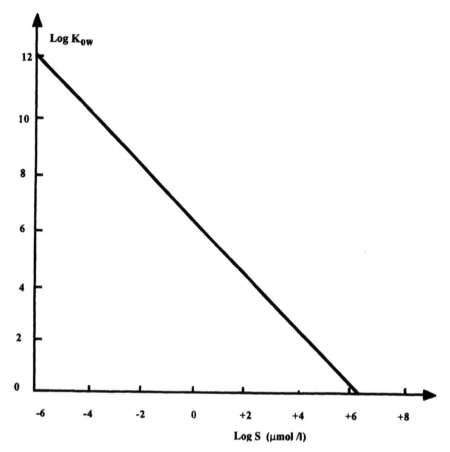

**FIGURE 11.1**
The relationship between log (water solubility in μmol/l) and log $K_{ow}$ at room temperature.

## 11.2 A Network of Estimation Methods

Estimation methods for the physical–chemical properties of chemical compounds were applied 40 to 60 years ago, as they were urgently needed in chemical engineering. They are, to a high extent, based on contributions to a focal property by molecular groups and the molecular weight, the boiling point, the melting point, and the vapor pressure as a function of the temperature. In addition a number of auxiliary properties result from these estimation methods, such as the critical data and the molecular volume. These properties may not have a direct application as ecotoxicological parameters in ecotoxicological models, but they are used as intermediate parameters, which may be used as the basis for estimation of other parameters.

The water solubility, the partition coefficient octanol–water, $K_{ow}$ and Henry's constant are crucial parameters in our network of estimation methods, because many other parameters are well correlated with these two. Three properties can be found for a number of compounds (or be estimated with a reasonably high accuracy) through knowledge of the chemical structure: the number of various elements, the number of rings, and the number of functional groups. In addition, there is a good relationship between water solubility and

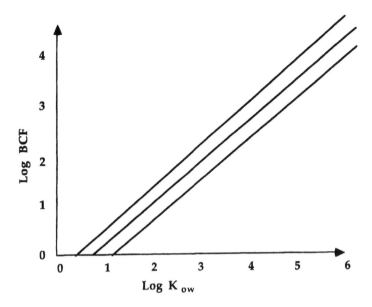

**FIGURE 11.2**
The relationship between log BCF and log $K_{ow}$ is shown. The BCF values are valid for fish of various species with a length of 20 to 30 cm. The correlation is based on several sources, with a total of 142 observations. The diagonal lines indicate the band corresponding to 95% of the observations on which the correlation is based.

$K_{ow}$ (see Figure 11.1). In the last decade many good estimation methods for these three core properties have been developed.

During the last 20 years several correlation equations have been developed based on a relationship between the water solubility, $K_{ow}$ or Henry's constant on the one side, and physical, chemical, biological, and ecotoxicological parameters for chemical compounds on the other side. The most important of these parameters are: the adsorption isotherms soil–water, the rate of the chemical degradation processes (hydrolysis, photolysis, and chemical oxidation), the biological concentration factor (BCF), the ecological magnification factor (EMF), uptake rate, excretion rate, and a number of other ecotoxicological parameters. Both the ratio of concentrations in the sorbed phase and in water at equilibrium $K_a$ and BCF may often be estimated with relatively good accuracy by expressions such as $K_a$ or BCF = a log $K_{ow}$ + b. Numerous expressions with different a and b values have been published (see Jørgensen et al., 1991; Jørgensen, 1994). Figure 11.2 shows one of these relationships, in this case between BCF and log $K_{ow}$.

Many recent estimation methods are based on molecular connectivity, which measures the chemical structure by the use of a number called the molecular connectivity index. It would not be beneficial to go into detail to present the determination of molecular connectivity here, but those interested are referred to Jørgensen et al. (1997).

The biodegradation rate in water or soil is difficult to estimate, because the number of microorganisms varies several orders of magnitude from one type of aquatic ecosystem to the next and from one type of soil to the next. Artificial intelligence has been used as a promising tool to estimate this important parameter.

Several useful methods for estimation of biological properties are based on the similarity of chemical structures. The idea is that if we know the properties of one compound, it may be used to find the properties of similar compounds. If, for instance, we know the properties of phenol, which is named the parent compound, these may be used to give a more accurate estimation of the properties for monochlorophenol, dichlorophenol, trichlorophenol, and so on and for the corresponding cresol compounds. Estimation approaches based

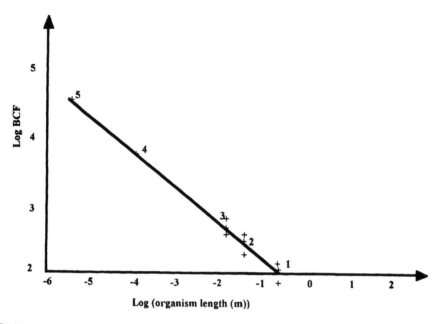

**FIGURE 11.3**

BCF for cadmium vs. size for various animals are shown. (1) goldfish (three different values are obtained), (2) mussels (three different values are obtained), (3) shrimp (32 different values are obtained), (4) zooplankton, (5) algae (based on three coinciding values).

on chemical similarity generally give a more accurate estimation, but they are also more cumbersome to apply because each estimation has a different starting point, namely the parent compound, with known properties.

Allometric estimation methods presume (Peters, 1983) that there is a relationship between the value of a biological parameter and the size of a considered organism. Usually it is presumed that the surface of the organism determines the value of the parameter because most parameters express some type of interaction between the organisms and the environment. An animal, for instance, with four times the surface would have four times the area of contact with the environment, and therefore would contain four times as much toxic substance. Since the volume increases eight times when the surface increases four times, this would mean that the concentration of the toxic substance would only be half as much. These methods make it possible to estimate important biological parameters for organisms of different size. Often, the length of the organism is used, which would give a linear log–log relationship with the slope +2 for rates, while it would give a slope of –1 for specific rates and concentrations. Figure 11.3 shows an example. Log BCF is plotted vs. log (organism length) in m (Jørgensen, 1984). The relationship is a straight line with a slope of –1, as expected.

The ecotoxicological parameters, $LC_{50}$, $LD_{50}$, MAC, EC, and NEC, can be estimated from a wide spectrum of physical and chemical parameters, although these estimation equations are generally more inaccurate than the estimation methods for physical, chemical, and biological parameters. Molecular connectivity or chemical similarity for estimation of toxicological parameters usually offer better accuracy.

Jørgensen et al. (1997) have developed a software named WINTOX, based on a network of the above-mentioned estimation methods. The software is easy to use and can rapidly provide estimations. Each relationship between two properties is based on the average result obtained from a number of different equations found in the literature. There is, however, a price to pay for using such an "easy" software. The accuracy of the estimations is not as good as the more sophisticated methods based on similarity in chemical structure, but in many, particularly modeling, contexts the results found by WINTOX offer sufficient accuracy. In addition, it is always useful to come up with a first intermediate guess.

The software makes it possible to start the estimations from the properties of the chemical compound already known. The accuracy of the estimation using this software can be improved considerably by knowing a few key parameters (for instance, the boiling point and Henry's constant). As it is possible to get software which is able to estimate Henry's constant and $K_{ow}$ with a higher accuracy than WINTOX can, combining these software packages with WINTOX is recommended. Another possibility would be to estimate a couple of key properties by use of chemical similarity methods and then use these estimations as known values in WINTOX. These methods for improving the accuracy will be discussed in the next section.

## 11.3   Estimation: A Trade-Off Between Complexity and Accuracy

The various estimation methods may be classified into two groups:

A.  General estimation methods based on an equation of general validity for all types of compounds, even though some of the constants may depend on the type of chemical compound, or they may be calculated by adding contributions (increments) based on chemical groups and bonds.

B.  Estimation methods valid for a specific type of chemical compounds, for example aromatic amines, phenols, aliphatic hydrocarbons, and so on. The property of at least one key compound is known. Based on the structural differences of the key compound from all other compounds of the considered type, (for instance, two chlorine atoms are substituted for hydrogen in phenol to get 2,3-dichlorophenol) and the correlation between the structural differences and the differences in the considered property, the properties for all compounds of the considered type can be found. These methods are based on chemical similarity.

Methods of class B are generally more accurate than methods of class A, but they are more cumbersome to use because the right correlation must be found for each property of each chemical. Furthermore, the requested properties should be known for at least one key component, which may be difficult when a series of properties are needed. If estimation of the properties for a series of compounds belonging to the same chemical class is required, it is tempting to use a suitable collection of class B methods.

Methods of class A form a network which facilitates the possibilities of linking the estimation methods together in a computer software, (e.g., WINTOX). The network of WINTOX as an example of these estimation networks is illustrated in Figure 11.4. As it is a network of class A methods, it should not be expected that the accuracy of the estimations is as high as can be obtained by the more specific class B methods. With WINTOX it is, however, possible to estimate the most pertinent properties directly from the structural formula.

WINTOX is based on average values of results obtained by simultaneous use of several estimation methods for most of the parameters. It implies increased accuracy of the estimation,

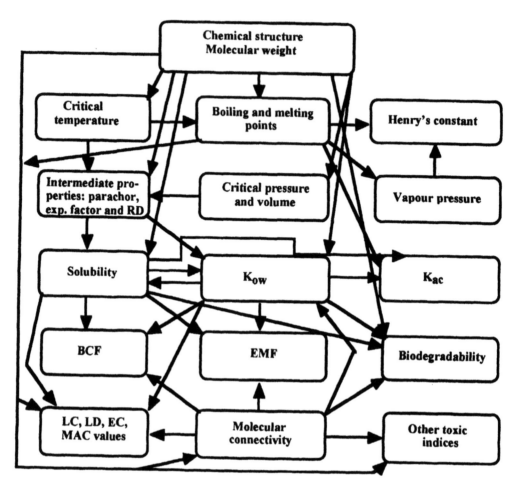

**FIGURE 11.4**
The network of estimation methods in WINTOX is shown. An arrow represents a relationship between two or more properties.

mainly because it gives a reasonable accuracy for a wider range of compounds. If several methods are used in parallel, a simple average of the parallel results has been used in some cases, while a weighted average is used in other cases where it has been found beneficial for the overall accuracy of the program. When parallel estimation methods give the highest accuracy for *different* classes of compounds, use of weighting factors seems to offer a clear advantage.

It is generally recommended that as many estimation methods as possible be applied for a given case study to increase the overall accuracy. If the estimation obtained by WINTOX can be supplied by other recommendable estimation methods, it can be strongly recommended.

It is also possible, to achieve a higher accuracy if some of the properties are known in addition to the structural formula. Particularly, if the boiling point, Henry's constant, and/or the partition coefficient octanol–water are known, the accuracy will increase significantly, even for the toxicological properties (parameters).

The accuracy obtained by the software can therefore be increased relatively easily by combining the use of the software with databases and estimation methods, which are readily available, for Henry's constant and the partition octanol–water coefficient. The two software programs are:

Henry's Law Constant Program
William M. Meylan
Philip H. Howard
CRC/Lewis Publishers, Boca Raton, FL 33431 U.S.

CLogP for Windows
BioByte Corp.
201 W. Fourth St., Suite #204
Claremont, CA 91711, U.S.

These two programs are based on a very large range of compounds and can, for these two properties, offer a higher accuracy than WINTOX. Moreover, it should be recommended to apply Beilstein to find such common physical data as boiling and melting points.

It is further recommended to apply other available, additional estimation methods. They may at least be used to confirm results obtained by WINTOX. Additional estimation methods may be found in the many references to QSAR-methods including: Jørgensen et al., 1997; Lyman, W.J., Reehl, W.F., and Rosenblatt, D.H., 1990.

This does not imply that we can reach the accuracies that characterize the class B methods, but that we can always obtain accuracies that make it advantageous to use this software, as it is much more rapid to apply than a suitable collection of class B methods, which in most cases have to be found by a comprehensive literature search.

It is not possible to give general recommendations on which methods or combinations of methods to use in a given situation. It may be desirable to know some properties with a particularly high accuracy, while it is sufficient to know other properties within coarse ranges. It is therefore crucial in each situation to select the right combination of estimation methods which will meet the demands for a defined accuracy most rapidly. It is necessary to find the trade-off between very complex, time-consuming, but accurate methods and simple, rapid, more inaccurate methods. Here may the use of WINTOX, in combination with the databases and software for Henry's constant and the partition octanol–water coefficient, offer an attractive solution which is relatively rapid and with medium accuracy.

WINTOX does not comprise estimations of biodegradability, because this parameter cannot be estimated with fully acceptable accuracy. In addition, biodegradation depends on a number of environmental factors, of which some can be quantified and others cannot. The number of estimation methods for biodegradation rates is limited compared with many other properties of environmental interest due to these limitations. Some estimation methods for biodegradation rates are given in the literature, and it can of course be recommended to apply them, but the results should be considered a first coarse approximation and should only be used as a preliminary value. The highest accuracy is obtained for the estimation methods that focus on the biodegradation in activated sludge plants because the conditions are more similar among activated sludge plants than among wetlands or lakes, for example. Particularly for biodegradation rates, an actual determination in the laboratory (and even better in a microcosm or *in situ*) of this property is recommended. Biodegradation is often a very sensitive parameter in ecotoxicological models, particularly for very mobile compounds which emphasizes the importance of a more accurate determination of this property.

WINTOX can also be recommended as an educational tool for courses in environmental chemistry and ecotoxicology. Due to its relatively fast generation of estimations, it will be possible to attain within given time limits several estimations, including very illustrative comparisons of properties of several compounds which opens up discussion of the different environmental consequences of application of various chemicals. When WINTOX is used as an educational tool, its accuracy/inaccuracy should be illustrated by estimating at least a few parameters which are already known from the literature. A discussion of the discrepancies between the estimated and known values should take place, because that would inevitably lead to a discussion of the use of safety factors and that we have no other choice than to estimate or determine a parameter which cannot be found in the literature. As the determination is very time consuming, the choice in practice is most often reduced to the use of estimations.

## 11.4   Summary and Conclusions

Many methods for estimation of the properties of chemical compounds have been developed. The various estimation methods may even be linked in a network which makes it possible to estimate a wide range of parameters (properties), provided that the chemical structure is known. WINTOX is mentioned as an example of such a network of estimation methods (see also Figure 11.4). The accuracy of the estimation is improved considerably if one or more properties are known from the literature or from other more accurate but perhaps more time-consuming estimation methods.

Estimation methods are urgently needed in environmental risk assessment and other environmental management contexts, because our knowledge of the properties and effects of toxic substances in nature is very limited. The standard deviation for estimated parameters is, of course, higher, sometimes much higher, than for measured values, but by a proper use of a combination of estimation methods, it is possible to make estimations with an acceptably low standard deviation.

## References

Jørgensen, S.E., 1984, Parameter estimation in toxic substance models, *Ecological Modelling*, 22, 1.

Jørgensen, S.E., 1990, *Modelling in Ecotoxicology*, Elsevier, Amsterdam.

Jørgensen, S.E., 1991, *Modelling in Environmental Chemistry, Developments in Environmental Modelling,* 17, Elsevier, Amsterdam.

Jørgensen, S.E., 1994, *Fundamentals of Ecological Modelling,* 2nd ed., *Developments in Environmental Modelling,* 19, Elsevier, Amsterdam.

Jørgensen, S.E., 1995, Exergy and ecological buffer capacities as measures of ecosystem health, *Ecosystem Health*, 1, 150.

Jørgensen, S.E., 1997, *Integration of Ecosystem Theory: A Pattern*, Kluwer, Dordrecht, The Netherlands.

Jørgensen, S.E. and Johnsen, I., 1989, *Principles of Environmental Science and Technology, Studies in Environmental Science*, 33, Elsevier, Amsterdam.

Jørgensen, S.E., Nielsen, S.N., and Jørgensen, L.A., 1991, *Handbook of Ecological Parameters and Ecotoxicology*, Elsevier, Amsterdam.

Jørgensen, S.E., Halling-Sørensen, B., and Nielsen, S.N., 1995, *Handbook of Environmental and Ecological Modeling*, CRC Lewis Publishers, Boca Raton.

Jørgensen, S.E., Mahler, H., and Halling Sørensen, B., 1997, *Handbook of Estimation Methods in Environmental Chemistry and Ecotoxicology*, Lewis Publishers, Boca Raton.

Lyman, W.J., Reehl, W.F., and Rosenblatt, D.H., 1990, *Handbook of Chemical Property Estimation Methods. Environmental Behaviour of Organic Compounds*, American Chemical Society, Washington, D.C.

Mackay, D., 1991, *Multimedia Environmental Models. The Fugacity Approach*, Lewis Publishers, Boca Raton.

Mackay, D. and Paterson, S., 1982, Fugacity revisited, *Environ. Sci. Technol.*, 16, 654A.

Peters, R.M., 1983, *The Ecological Implications of Body Size*, Cambridge University Press, Cambridge, U.K.

Wolfram, S., 1984a, Cellular automata as models of complexity, *Nature*, 311, 419.

Wolfram, S., 1984b, Computer software in science and mathematics, *Sci. Am.*, 251, 140.

# 12

## Ecological Engineering

**S.E. Jørgensen and William Mitsch**

## CONTENTS

## 12.1 What Is Ecological Engineering?

H.T. Odum was among the first to define ecological engineering (Odum, 1962; Odum et al., 1963) as the "environmental manipulation by man using small amounts of supplementary energy to control systems in which the main energy drives are coming from natural sources." Odum further developed the concept (Odum, 1983) of ecological engineering as follows: "ecological engineering, the engineering of new ecosystems designs, is a field that uses systems that are mainly self-organizing."

Straskraba (1984, 1985) has defined ecological engineering, or as he calls it, ecotechnology, more broadly as the use of technological means for ecosystem management, based on a deep ecological understanding, to minimize the costs of measures and their harm to the environment. Ecological engineering and ecotechnology are considered to be different by Straskraba but are considered synonymous by many others.

Mitsch and Jørgensen (1989) give a slightly different definition which, however, covers the same basic concept as the definition given by Straskraba and also encompasses the definition given by H.T. Odum. They define ecological engineering and ecotechnology as the design of human society with its natural environment for the benefit of both. It is engineering in the sense that it involves the design of man-made or natural ecosystems or parts of ecosystems. It is, as all engineering disciplines, based on basic science, in this case ecology and system ecology. The biological species are the components applied in ecological engineering. Ecological engineering represents, therefore, a clear application of ecosystem theory.

At a May 1993 workshop on ecological engineering at the U.S. National Academy of Sciences (see Mitsch, 1996, 1998), a slight variation of the definition of ecological engineering, as originally given by Mitsch and Jørgensen (1989), was presented: "The design of sustainable ecosystems that integrate human society with its natural environment for the benefit of both."

Ecotechnic is another often applied word, but in addition to ecotechnology or ecological engineering it encompasses the development of technology applied in society, based on ecological principles, for instance, all technologies based upon cycling or the use of resources in a more environmentally friendly way.

Ecological engineering should furthermore not be confused with bioengineering or biotechnology (Mitsch and Jørgensen, 1989; Mitsch, 1993). Biotechnology involves the manipulation of the genetic structure of the cells to produce new organisms capable of carrying certain functions. Ecotechnology does not manipulate at the genetic level, but at several steps higher in the ecological hierarchy. The manipulation takes place on an assemblage of species or their abiotic environment as a self-designing system that can adapt to changes brought about by outside forces, controlled by human or by natural forcing functions.

Ecological engineering is also not the same as environmental engineering, which is involved in cleaning up processes to prevent pollution problems. It uses settling tanks, filters, scrubbers, and man-made components which have nothing to do with the biological and ecological components that are applied in ecological engineering, although the use of environmental engineering also aims to reduce man-made forcing functions on ecosystems. Ecotechnic may be considered to include, in addition to ecological engineering, environmental technology based on ecological principles such as recirculation and a better use of the resources.

The tool box of the two types of engineering are completely different. Ecological engineering uses ecosystems, communities, organisms, and their immediate abiotic environment.

All application of technologies is based on quantification. Ecosystems are very complex, and the quantification of their reactions to impacts or manipulations therefore becomes complex. Fortunately, ecological modeling represents a well-developed tool to survey ecosystems, their reactions, and the linkage of their components. Ecological modeling is able to synthesize our knowledge about an ecosystem and makes it possible to a certain extent to quantify any changes in ecosystems resulting from the use of both environmental engineering and ecological engineering.

Ecological engineering may also be used directly to design constructed ecosystems. Consequently, ecological modeling and ecological engineering are two closely cooperating fields. Research in ecological engineering was originally covered by the journal *Ecological Modelling*, which has been renamed *Ecological Modelling — International Journal on Ecological Modelling and Engineering and Systems Ecology* to emphasize the close relationship between the three fields: ecological modeling, ecological engineering and systems ecology. *Ecological Engineering* was launched as an independent journal in 1992 and has successfully covered the field of ecological engineering which has grown rapidly during the 1990s due to the increasing acknowledgment of the need for technologies other than environmental technology in our effort to solve pollution problems. This development does not imply that ecological modeling and ecological engineering are moving in different directions. On the contrary, ecological engineering is increasingly using models to perform design of constructed ecosystems or to quantify the results of application of specific ecological engineering methods for comparison with alternative methods.

## 12.2    Examples and Classification of Ecological Engineering

Ecotechnology may be based on one or more of the following four classes of ecotechnology:

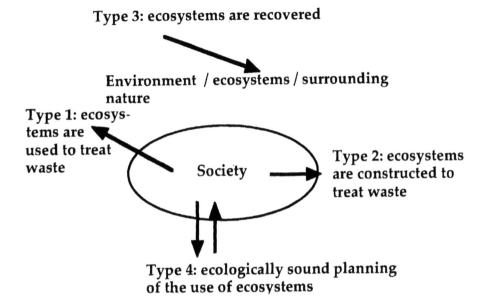

**FIGURE 12.1**
An illustration of the four types of ecological engineering.

1. Ecosystems are used to reduce or solve a pollution problem that otherwise would be (more) harmful to other ecosystems. A typical example is the use of wetlands for wastewater treatment.

2. Ecosystems are imitated or copied to reduce or solve a pollution problem, leading to constructed ecosystems. Examples are fishponds and constructed wetlands for treatment of wastewater or diffuse pollution sources.

3. The recovery of ecosystems after significant disturbances. Examples are coal mine reclamation and restoration of lakes and rivers.

4. The use of ecosystems for the benefit of humankind without destroying the ecological balance, i.e., utilization of ecosystems on an ecologically sound basis. Typical examples are use-integrated agriculture and development of organic agriculture. These types of ecotechnology find wide application in the ecological management of renewable resources.

The idea behind these four classes of ecotechnology is illustrated in Figure 12.1. Notice that ecotechnology is operating in the environment/the ecosystems. It is here ecological engineering has its tool box.

All four classes of ecological engineering may find illustrative examples when ecological engineering is applied to replace environmental engineering because ecological engineering methods offer an ecologically more acceptable solution and when ecological engineering is the only method that can offer a proper solution to the problem. These examples are shown in Table 12.1, in which the alternative environmental technological solution is also indicated.

It does not imply that ecological engineering will consequently replace environmental engineering. On the contrary, the two technologies should work hand in hand to solve environmental management problems better than either could do alone. This is illustrated in Figure 12.2, where a proper control of lake eutrophication requires both ecological engineering and environmental technology.

**TABLE 12.1**

Ecological Engineering Examples
(alternative environmental engineering methods are given, when possible)

| Type of Ecological Engineering | Example of Ecological Engineering | | Environmental Engineering Alternative |
|---|---|---|---|
| | Without Environmental Engineering Alternative | With Environmental Engineering Alternative | |
| 1 | Wetlands utilized to reduce diffuse pollution | Sludge disposal on agricultural land | Sludge incineration |
| 2 | Constructed wetland to reduce diffuse pollution | Root zone plant | Traditional wastewater treatment |
| 3 | Recovery of lakes | Recovery of contaminated land *in situ* | Transport and treatment of contaminated soil |
| 4 | Agro forestry | Ecologically sound planning of harvest rates of resources | |

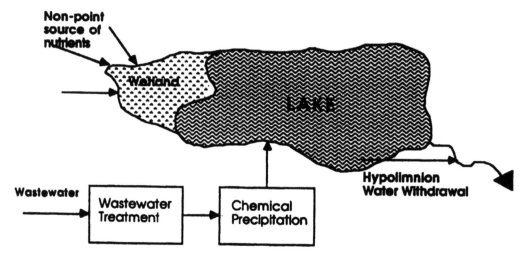

**FIGURE 12.2**

Control of lake eutrophication with a combination of chemical precipitation for phosphorus removal from wastewater (environmental technology), a wetland to remove nutrients from the inflow (ecotechnology, type 1 or 2), and siphoning off hypolimnetic water, rich in nutrients, downstream (ecotechnology, type 3).

Type 1 ecological engineering, application of ecosystems to reduce or solve pollution problems, may be illustrated by wetlands utilized to reduce the diffuse nutrient loadings of lakes. This problem could not be solved by environmental technology. Treatment of sludge could be solved by environmental technology, namely by incineration, but the ecological engineering solution, sludge disposal on agricultural land, which implies a utilization of the organic material and nutrients in the sludge, is a considerably more sound method from ecological perspectives.

The application of constructed wetlands to cope with diffuse pollution is a good example of type 2 ecological engineering. Again, this problem cannot be solved by environmental technology. The application of root zone plants for treatment of a small amount of wastewater is an example of ecological engineering, type 2, where the environmental engineering alternative, a mechanical–biological–chemical treatment cannot compete, because it will be too costly relative to the amount of wastewater (sewage system, pumping stations, and so on). A solution requiring fewer resources will always be more sound ecologically.

Recovery of land contaminated by toxic chemicals is possible by environmental technology, but it requires transportation of the soil to a soil treatment plant, where biological biodegradation of the contaminants takes place. Ecological engineering proposes a treatment *in situ* by adapted microorganisms or plants. The latter method is much more cost moderate, and the pollution related to the transport of soil will be omitted. Recovery of lakes by biomanipulation, installation of an impoundment, sediment removal or coverage, siphoning off hypolimnetic water, rich in nutrients, downstream, or by several other proposed ecological engineering techniques are other type 3 examples of ecological engineering. It is hardly possible to obtain the same results by environmental engineering because it requires activities in the lake and in the vicinity of the lake.

Type 4 ecological engineering is largely based on prevention of pollution by utilization of ecosystems on an ecologically sound basis. It is hardly possible to find environmental engineering alternatives in this case, but it is clear that a prudent harvest rate of renewable resources, whether it is timber or fish, is the best long-term strategy from an ecological and economic point of view. An ecologically sound planning of the landscape is another example of the use of class 4 ecological engineering.

## 12.3  Ecosystem Theory Applied in Ecological Engineering

Ecological engineering has been presented in the two previous sections of this chapter as a useful technological discipline. This section is devoted to presenting a number of ecosystem principles ecological engineering methods in practice.

Mitsch and Jørgensen (1989) apply 12 system ecological principles to understand the basic concepts of ecotechnology and to ensure an ecologically sound application. All 12 principles are based on system concepts applied in ecology. The principles are presented below with at least two examples as illustrations of the application and consideration of each of the 12 in the practical use of ecological engineering. The two examples for each principle are taken from terrestrial ecosystems, often from agriculture, and from aquatic ecosystems, respectively.

**Principle 1: Ecosystem structure and functions are determined by the forcing functions of the system.** Ecosystems are open systems, which implies that they exchange mass and energy with the environment. There is a close relationship between the anthropogenic forcing functions and the state of agricultural ecosystems, but due to the openness of all ecosystems, the adjacent ecosystems are also effected. Intensive agriculture leads to drainage of the surplus nutrients and pesticides to the adjacent ecosystems. This is the so-called nonpoint or diffuse pollution. The abatement of this source of pollution requires a wide use of ecotechnological methods.

The use of constructed wetlands and impoundments to reduce the concentrations of nutrients in streams entering a lake ecosystem illustrates the application of this system ecological principle in ecotechnology. The forcing functions, the nutrient loadings, are reduced and a corresponding reduction of the eutrophication should be expected.

**Principle 2: Homeostasis of ecosystems requires accordance between biological function and chemical composition.** The biochemical functions of living organisms define their composition, although these are not to be considered as fixed concentrations, but rather as ranges. The application of the principle implies that the flows of elements through agricultural systems should be according to the biochemical stoichiometry. If this is not the case, the elements in surplus will be exported to the adjacent ecosystems and make an impact on the natural balances and processes there. An investigation (R. Skaarup and T. Sørensen, 1994) of a well-managed Danish farm has shown that much can be gained by a complete material flow analysis. The results will not only lead to reduction in the emission level of pollutants but will often also imply cost reductions.

Several restoration projects on lakes have considered these principles. The limiting nutrient determining the eutrophication of a lake is found, and the selected restoration method will reduce the limiting nutrient further. When sediment is removed or covered in lakes, it is to prevent the otherwise limiting nutrient, phosphorus, to reach the water phase.

**Principle 3: It is necessary in environmental management to match recycling pathways and rates to ecosystems to reduce the effect of pollution.** The application of sludge as a soil conditioner illustrates this principle very clearly. The recycling rate of nutrients in agriculture has to be accounted for in any use of sludge. If the sludge is applied faster than it can be utilized by the plants, a significant amount of the nutrients might contaminate the streams, lakes, and groundwater adjacent to the agricultural ecosystem. If, however, the influence of temperature on the nitrification and denitrification, the hydraulic conductivity of the soil, the slope of field, and the rate of the plant growth all are considered in an application plan for the manure, the loss of nutrients to the environment will be maintained at a very low and probably acceptable level. This is possible by ecological models to develop a plan for the application of the sludge.

Elements are recycled in agrosystems but to a far lesser extent than occurs in nature. For the last couple of decades animal husbandry has become separated from plant production. This has made internal cycling more difficult to achieve, of less economical value, and thus less attractive. We have come to accept losses of these relatively inexpensive, easy-to-apply artificial fertilizers, and have compensated by increasing their application.

So, the message is: know the ecological processes of farming and their rates and manage the system accordingly, i.e., recycle as much as possible at the right rates and don't use more fertilizer than can be recycled.

The recovery of a eutrofe lake by application of shading is an illustrative example of the same principle for aquatic ecosystems.

Ecosystems with pulsing patterns often have greater biological activity and chemical cycling than systems with relatively constant patterns. A specific case study will illustrate the recognition of pulsing force and how it is possible to take advantage of it in ecological engineering. Figure 12.3 shows a map of an estuary in Brazil named Cannaneia. The shore of the islands and the coast are very productive mangrove wetlands, and the entire estuary is an important nesting area for fish and shrimp. Channel C (see the map) was built to avoid flooding upstream, where productive agricultural land is situated. The construction of the channel has caused a conflict between farmers, who want the channel open, and fishermen, who want it closed due to its reduction of the salinity in the estuary (the right salinity is of great importance for mangrove wetlands). The estuary is exposed to tide, which is important for the maintenance of good water quality with a certain minimum of salinity. The conflict can be solved by use of an ecological engineering approach that takes advantage of the pulsing force (the tide). A sluice in the channel could be constructed to discharge the fresh water when it is most appropriate. The tide would, in this case, be used to transport the fresh water as rapidly as possible to the sea. The sluice should be closed when the tide is on its way into the estuary. The tidal pulse frequently is selectively filtered to produce an optimal management situation.

**Principle 4: Ecosystems are self-designing systems. The more one works with the self-designing ability of nature, the lower the costs of energy to maintain that system.** Many of our actions are undertaken to circumvent or to counteract the process of self-design. For example, the biodiversity of agricultural fields would be significantly higher if pesticides were not used and nature left to rule on its own. While the self-designing systems are able

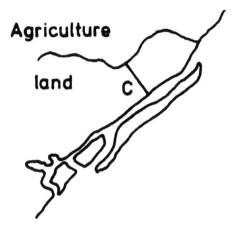

**Agriculture land** C

**FIGURE 12.3**
Map of Cannaneia Estuary in Brazil. Channel C was built to avoid flooding agricultural land upstream.

to implement sophisticated regulations before violent fluctuations or even chaotic events occur, agriculture attempts to regulate chemically, e.g., killing undesirable organisms with pesticides. This very coarse regulation sometimes does more harm than anticipated, for instance when the insects' predators are affected more than the insects. The conclusion seems clear: don't eliminate natural regulation mechanisms, i.e., maintain a pattern of nature within agricultural systems.

The application of green fields during the winter in northern Europe is consistent with this principle, as the self-designing ability is maintained of this time of the year. Bare soil should generally be avoided because of possible erosion.

The closer the agricultural system is to a natural ecosystem, the more self-designing capacity it has. An integrated agriculture system is therefore less vulnerable than modern industrialized agriculture, because it offers more components for self-designing regulations and it has a wider range of flows that facilitate the possibilities for recycling.

The use of constructed wetlands in lake restoration is an example of the application of this principle of self-design taken from aquatic ecosystems. If we design a wetland to partially remove nutrients from streams entering a lake, the lake itself can do the self-design and reduce the level of eutrophication accordingly. The constructed wetland will also use self-design. The diversity (complexity) and nutrient removal efficiency will increase gradually, provided the wetland is undisturbed.

**Principle 5: Processes of ecosystems have a characteristic time and space scale, that should be accounted for in environmental management.** Environmental management should consider the role of a certain spatial pattern for the maintenance of biodiversity. Violation of this principle by drainage of wetlands and deforestation on too large a scale has caused desertification. Wetlands and forests maintain high humidity in the soil and regulate the precipitation. When the vegetation is removed, the soil is exposed to direct solar radiation, and it dries, causing organic matter to be burned off. Use of fields that are too large prevents wild animals and plants from finding their ecological niches, an important component in the pattern of agriculture and more or less untouched nature. The conclusion is to maintain ditches and hedgerows as corridors in the landscape or as ecotones between agricultural and other ecosystems. Fallow fields also should be planned as contributors to the pattern of the landscape.

The example mentioned above on the use of the tide to transport fresh water as rapidly as possible to the sea may also be used to illustrate this principle of using the right time and space scale in the application of ecological engineering.

**Principle 6: Chemical and biological diversity contribute to the buffering capacity and the self-designing ability of ecosystems. A wide variety of chemical and biological components should be introduced or maintained for the ecosystem's self-designing ability to choose from. Thereby a wide spectrum of buffer capacities is available to meet the impacts from anthropogenic pollution.**
Biodiversity is important to buffer capacity and the ability of the system to meet a wide range of possible disturbances by the use of the ecosystem's self-designing ability. There are many different buffer capacities, corresponding to any combination of a forcing function and a state variable. It has been shown that vegetables cultivated as mixed cultures give a higher yield and are less vulnerable to disturbances (e.g., attacks by herbivorous insects). In agricultural practice this implies that it is advisable to use small fields with different crops.

This principle also implies that an integrated agriculture is less vulnerable than modern industrialized agriculture simply due to its higher biological and chemical diversity.

Recovery of lakes by use of biomanipulation usually increases the biodiversity and some buffer capacities.

**Principle 7: Ecotones, transition zones, are as important for ecosystems as membranes are for cells. Agricultural management should consider the importance of transition zones.** Nature has developed transition zones, denoted ecotones, to make a soft transition between two ecosystems. Ecotones may be considered buffer zones that are able to absorb undesirable changes imposed on an ecosystem from adjacent ecosystems. We must learn from nature and use the same concepts when we design interfaces between man-made ecosystems (agriculture, human settlements) and nature. Some countries have required a buffer zone between human settlements and the coasts of lakes or marine ecosystems (e.g., in Denmark it is 50 m).

Without buffer zones between agriculture and natural ecosystems emissions will be transferred directly to the ecosystem, while a buffer zone such as a wetland, would at least partly adsorb the emissions and thereby prevent their negative influences on natural ecosystems. Some countries require a buffer zone between arable land and streams or lakes (e.g., in Denmark it is 2 m).

A pattern of wetlands in the landscape can remove the emission of nitrates from agriculture, which is one of the hot issues in the environmental management of the diffuse pollution originating from agriculture.

The role of the littoral zone in lake management is another obvious example. A sound littoral zone with a dense vegetation of macrophytes will be able to absorb contamination before it reaches the lake.

**Principle 8: The coupling between ecosystems should be utilized to the benefit of ecosystems in the application of ecotechnology and in environmental management of agricultural systems.** An ecosystem cannot be isolated. It must be an open system, because it needs an input of energy to maintain the system (see also Principle 1). The coupling of agricultural systems with natural systems leads to transfer of pesticides and nutrients from agriculture to nature. Measures should be taken to achieve an (almost) complete utilization of pesticides and nutrients in the agricultural system, for instance by implementing proper fertilization plans accounting for these transfer processes. Ecological management should always consider all ecosystems as interconnected systems, not as isolated subsystems. It implies that not only local but also regional and global effects have to be considered. For instance, the methane emitted from rice fields may increase the greenhouse effect and thereby the global climate, which again will feed back to the cultivation of rice.

Lake management can only be successful if it is based on this principle, as all the inputs to the open ecosystem, a lake, should be considered. Recovery of a eutrofe lake will require that all nutrient sources are quantified and that a plan taking all the sources into account is realized.

**Principle 9: It is important that the application of ecotechnology and environmental management considers that the components of an ecosystem are interconnected, interrelated, and form a network, which implies that direct as well as indirect effects are important.** An ecosystem is an entity, or everything is linked to everything in the ecosystem. Any effect on any component in an ecosystem is therefore bound to have an effect on *all* components in the ecosystem either directly or indirectly, i.e., the entire ecosystem will be changed. It can be shown that the indirect effect is often more important than the direct one (Patten, 1982, 1991). Application of ecotechnology attempts to take the indirect effect into account, while management, which considers only the direct effects, often fails. There are numerous examples of the use of pesticides on herbivorous insects which also might have a pronounced effect on carnivorous insects and therefore will have an unintended effect on the herbivorous insects. Therefore, pesticides should not be used in a vacuum. Sufficient knowledge of the insect populations and their predators should be the basis for decision about the application of pesticides. Preferably, a model should be developed to assure that the pesticides reach the right target organisms and do not cause an adverse effect.

The use of ecotechnology in the abatement of toxic substances in aquatic ecosystems requires that this principle be considered. The biomagnification of toxic substances through the food chain is a result of the interconnectance of ecological components. Due to biomagnification, it is necessary to aim at a far lower concentration of toxic substances in aquatic ecosystems to avoid a undesirable high concentration of the toxic substance in fish for human consumption.

**Principle 10: It is important to realize that an ecosystem has a history in application of ecotechnology and environmental management in general.** The components of ecosystems have been selected to cope with the problems that nature has imposed on ecosystems for million of years. The high biodiversity of old ecosystems compared with immature ecosystems is another important realization of this principle. The structure of mature ecosystems should therefore be imitated in the application of ecological engineering. An ecosystem with a long history is better able to cope with emissions from its environment than an ecosystem with no history. This again emphasizes the importance of establishment of a pattern of agriculture and natural terrestrial and aquatic ecosystems to ensure that the history is preserved and the right solution can therefore be offered to emerging environmental problems.

Many lakes store considerable amounts of nutrients in their sediment due to a sad history of discharge of insufficiently treated wastewater. Restoration of such lakes often requires removal of the sediment, an extremely expensive restoration method for large and deep lakes. An ecologically sound management should take the history into account and use prevention at the proper time.

**Principle 11: Ecosystems are most vulnerable at the geographical edges. Therefore, ecological management should take advantage of ecosystems and their biota in their optimal geographical range.** When ecological engineering involves ecosystem manipulation, the system will have enhanced buffer capacity if the species are in the middle range of their environmental tolerance. The ecosystem manipulation should therefore consider a careful selection of the involved species in accordance with this principle. For agriculture it implies that the crops should be selected following this principle. The cultivation of tomatoes and other subtropical vegetables in northern Europe demonstrates how this principle is easily violated in modern agriculture. These products may compete in price due to good management or subsidies, but they cannot compete in quality.

Ecologically sound planning will use this principle and avoid the use of biological components which are at their geographical edges. This rule is important for both terrestrial and aquatic ecosystems.

**Principle 12: Ecosystems are hierarchical systems and all the components forming the various levels of the hierarchy make up a structure that is important for the function of the ecosystem.** It is significant to maintain the components that make up the landscape diversity, such as hedges, wetlands, shorelines, ecotones, ecological niches, etc. They will all contribute to the buffer capacity of the entire landscape. Clearly, integrated agriculture can follow this principle more easily than the industrialized agriculture, because it has more components to use for construction of a hierarchical structure.

It is equally important in our management of lakes to consider the benthic zone, the littoral zone, the epilimnion and hypolimnion. All the zones require the right conditions with respect to oxygen, pH, temperature, and so on to maintain the various organisms (the next lower level in the hierarchy) fitted to these zones. Selection of lake restoration methods requires consideration of this issue. Which zone has a problem maintaining the right ecological balance? What can we do to solve the problem? Which restoration methods should be selected? The answers to these questions require extensive use of ecological modeling, applied on ecotechnology.

## 12.4   Ecosystem Concepts in Ecological Engineering

Straskraba (1993) has presented seven principles of ecosystems which he has used to set up 17 rules on how to use ecotechnology. The rules may be expanded and applied to environmental management in general. They are based on basic scientific principles of systems ecology and should be respected whenever an environmental management decision should be taken. The principles (taken from Straskraba, 1993) are presented below with a short explanation, followed by the rules (Straskraba, 1993) with comments on their application in ecological technology and environmental management in general.

**Principle 1: Energy inputs to the ecosystems and available storage of matter are limited.** This principle is based on the conservation of matter and energy (Patten et al., 1997); nuclear processes may convert matter to energy, but these processes are insignificant in ecosystems.

The dominant energy input to life on Earth is solar energy. We may be able to supply this energy source with other forms of energy, for instance fossil fuel, but the only sustainable energy source is solar energy.

Favorable conditions for vegetation growth are represented by temperature, humidity (water), and sufficient concentration of nutrients. Mass conservation leads to the concept of limiting factors for growth and the Michaelis–Menten equation: growth = $v * S/(k + S)$, where S represents the concentration of the substrate or limiting nutrient; v and k are constants. A maximum growth rate is attained when $S \gg k$, which implies that another nutrient or factor becomes limiting.

**Principle 2: Information is stored in ecological systems in structures.** Structures are a result of the input of energy utilized to move away from thermodynamic equilibrium, gain exergy (see Chapters 2 through 4), or build structures. Such structures include organisms and physical structure of the landscape. Size is an important characteristic of structures. Organism size determines many important features of life, such as the rate of development, speed of movement, and the range of areas they inhabit. Certain minimum size of structures surrounding the organisms is necessary to satisfy their needs.

**Principle 3: The genetic code of single organisms is a storage of information about the development of the Earth's environment, about the process of adaptation to the changing**

**conditions of existence.** For organisms, information about past evolution is stored in the genes. The genetic code reflects not only the past of an individual organism, but also its surroundings, including other coexisting organisms. It is the consequence of tight coupling between each organism and its environment. The evolution of the organism is highly dependent on the condition of life (Straskraba, 1993).

In Chapters 3 and 4 it has been proposed to use calculations of exergy, denoted an exergy index, to describe the propensity for the development of an ecosystem and to assess ecosystem health. These proposals are in accordance with the concepts behind Principle 3.

**Principle 4: Ecosystems are open and dissipative systems. They are dependent on a steady input of energy from outside.** This principle is an application of the Second Law of Thermodynamics on ecosystems. An ecosystem must obey the conservation principle; it must also obey the basic scientific laws, e.g., the basic laws of thermodynamics. The energy input is utilized to cover the energy needs for maintenance, respiration, and evapotranspiration. If the energy (exergy) input exceeds these needs, the surplus energy (exergy) is utilized to build more structure.

**Principle 5: Ecosystems are multimediated feedback systems. Feedback systems are characterized by influences on one part of the system being fed back to other components of the system.** Feed backs create many completely unexpected effects, and the entire network describing the interrelations of the organisms may also work as feed backs. Patten (1991) has shown that indirect effects often may exceed direct effects due to the feed backs and the relationships determined by the network.

**Principle 6: Ecosystems have homeostatic capability, which results in smoothing out and depressing the effects of strongly variable inputs. However, this capability has a limited extent. Once exceeded, the system breaks down.** Several homeostatic mechanisms are known in biology, for instance the maintenance of pH in our blood and the maintenance of body temperature for warm-blooded animals. The homeostatic capability may be expressed by means of the concept of ecological buffer capacity (see Chapter 3). As the homeostatic capability is limited, so are buffer capacities. It is obviously important in environmental management to know and respect the buffer capacities. Otherwise the ecosystem may change radically and even collapse.

**Principle 7: Ecosystems are adaptive and self-organizing systems.** We may distinguish between biochemical and biological adaptation. Biochemical adaptation takes place when the individual organism adapts to changed conditions. Phytoplankton, for instance, can adjust its content of chlorophyll a to the intensity of solar radiation. Biological adaptation implies adaptation on the ecosystem level. The properties of the species in the ecosystem are changed in accordance with prevailing conditions, either for the next generation by dominant heritage of the properties best fitted to the prevailing conditions, or by complete or partial replacement of species with different tolerances. Other species waiting in the wings take over, if they have a combination of properties better fitted to the (new) emergent conditions.

Self-organization of the ecosystems may be understood as a directional change of the species composition of the ecosystem, when its surroundings change, e.g., by human activities (Straskraba, 1993).

The principles and their immediate ecotechnological implications are summarized in Figure 12.4.

These principles give rise to 17 rules (Straskraba, 1993) for the application of ecotechnology or environmental management, in general. The 17 rules with reference to the principles are presented below.

**Rule 1: Minimize energy waste.** This rule is based on the conservation principle (see Principle 1). Energy is limited and therefore a resource. In addition, all uses of energy cause pollution: carbon dioxide for fossil fuel, nuclear waste for nuclear energy, noise and landscape

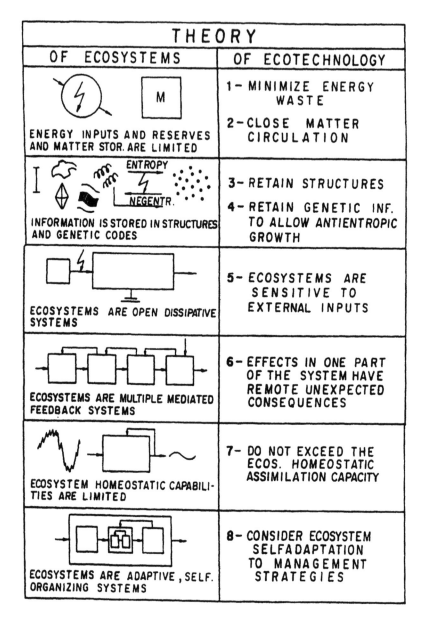

**FIGURE 12.4**

Theoretical principles of ecosystems and their reflection in the theory and application of ecotechnology. (From Straskraba, M., 1993, Ecotechnology as a new mean for environmental management, in *Ecological Engineering*, 2, 4, 311. With permission.)

pollution for windmills, etc. The selection of environmental technology should also be based on this rule and rule number two, because that would mean that not only are we reducing the use of resources, but we are also reducing pollution correspondingly, as energy not wasted and matter recycled are not discharged to the environment.

**Rule 2: Recycle.** This rule is also based on the conservation principle, Principle 1. It is a core principle in cleaner technology (see Chapter 6) and is an imitation of ecosystems. Ecosystems use, for all essential elements, recycling, which will ensure that the resources can be used again and again. The development of green audits is based on the conservation principle. The purpose is to increase recycling and reduce waste (see Chapters 5 and 6).

**Rule 3: Retain all kinds of structures.** This rule is based on Principle 2. It is in accordance with the use of ecotechnology, which prescribes that hedges, littoral zones, a landscape pattern of agriculture, wetlands, forests, hedges, and hills are essential.

**Rule 4: Consider long-term horizons.** Sustainable development, or sustainable life, can be realized only when we switch from short-term evaluation of our economic practices to long-term horizons (Straskraba, 1993). This implies that we need to consider the long-term consequences of changes in the ecological structures and extinction of species due to our activities (Principles 2 and 3).

**Rule 5: Do not neglect that mankind is dependent on many organisms and that their loss may lessen our ability to survive in the steadily changing environment.** This rule is based on Principle 3. Our dependence on many species is not necessarily completely known today. We may find medicine which can cure cancer in a rainforest plant which has not yet been discovered. The presence of important indirect effects makes it very difficult to evaluate the importance of many species. It seems, therefore, to be a prudent strategy to maintain as many species as possible and maintain the networks in which they operate to the highest possible extent.

**Rule 6: Consider ecosystem dynamics.** This rule is directly translated from Principle 4. Open and dissipative systems have complex dynamics, which should not be neglected in environmental practices. Dynamic and structurally dynamic models appear to be tools needed for decisions concerning ecosystem management, as they consider these important dynamics of ecosystems.

**Rule 7: Understand that nature is a teaching ground for handling complex systems. In particular, the ability to survive by adapting to changing conditions can be learned only from nature.** This rule is based on Principles 4 and 5. The study of adaptation and evolution of individual organisms and ecosystems seems to be the only way we can recognize our own possibilities and limitations of survival under conditions of the environment subject to unexpected major changes created by our activity. The recent development of cleaner technology and organic agriculture is based on imitation of nature or learning from nature.

**Rule 8: People, as part of the ecosystem, are dependent on solar energy directly. We need it in the form of light to see and in the form of heat for warmth. We are also indirectly dependent on energy stored in fossil fuel as well as in vegetable and animal food. We are dissipative structures, too.** This rule acknowledges that Principle 4 is also valid for us — we are open and dissipative systems, and we are parts of open and dissipative systems. We are therefore also victims of Principle 4.

**Rule 9: Understand that we are dependent on and sensitive to external inputs, in particular solar radiation, but also to the supply of minerals from the Earth.** This rule is based upon Principles 1, 2, and 4. It is also associated with rules 1, 2, and 8. Our activity should not alter the inputs by solar radiation through destruction of the ozone layer because it may cause skin cancer. Supplies of minerals and energy are needed because we are indeed a part of the global dissipative structure. Therefore, we should minimize energy waste and recycle.

**Rule 10: Manage the environment as an interconnected system, not as isolated subsystems.** This rule is also based on Principle 4. You cannot close an ecosystem. It has to be open, which means it has inputs and outputs, but outputs from one ecosystem may be inputs for another ecosystem. Pollution problems can therefore not be solved by discharge/emission/dilution, but only by degradation and recycling.

**Rule 11: Evaluate available management options simultaneously.** Any technological goal can be achieved in different ways, using different options. The options are, however, different in expenditures, quality of product, material needs, waste, etc. The options available to solve an environmental problem have to be recognized and evaluated with respect to best satisfying the goal while respecting all the principles and rules presented in this section.

**Rule 12: Include secondary effects elsewhere.** This rule is based on Principle 5, and partly Principle 6. It is difficult to survey the consequences of our activities for ecosystems due to the complexity and openness of these systems and due to the often surprising indirect effect. As wide an evaluation as possible — if possible a global system evaluation — of the consequences of our activities is therefore needed.

**Rule 13: Do not exceed the ecosystem homeostatic capacity.** This rule is directly derived from Principle 6. It is important to know the buffer capacities of focal ecosystems (see Chapter 3), possibly by use of models to consider the indirect effects and the ability of ecological networks to absorb disturbances.

**Rule 14: Consider the self-adaptation to management strategies.** This rule is directly based on Principle 7. The self-adaptation ability of natural components in ecosystems should be reflected in our environmental management strategy. The best illustration may be taken from pest control. The use of pesticides is problematic because the organisms become adapted to the pesticides, while biological methods do not violate Principle 7. Unfortunately, there are numerous examples of environmental management where we have omitted this rule and therefore failed. To mention a few: introduction of rabbits to Australia and introduction of the Nile Perch to Lake Victoria.

**Rule 15: Evaluate the socioeconomic environment.** This rule is an acknowledgment of the self-organization (Principle 7) of the human society, which is another complex, open, dissipative system with many properties similar to an ecosystem's. As the society and ecosystems both are open, they are also interchanging matter and energy. The ultimate environmental management should therefore consider both types of system which makes the problem even more complex. It has been attempted to develop ecological–sociological–economic models, but this type of model is in its very early infancy.

**Rule 16: Evaluate all possible human uses of the environment.** The same environmental area or object, like a body of water or a forest, can be used for different purposes. When decisions are made, all of these possible purposes have to be taken into account to find the most environmentally friendly use.

**Rule 17: Base measures on ecosystem principles.** In other words, use the principles presented in this section and in Section 12.3.

---

## References

Mitsch, W.J., 1993, Ecological engineering — a cooperative role with the planetary life-support systems, *Environmental Science & Technology*, 27, 438.

Mitsch, W.J., 1996, Ecological engineering: a new paradigm for engineers and ecologists, in Schulze, P.C., Ed., *Engineering within Ecological Constraints*. National Academy Press, Washington, D.C., 111.

Mitsch, W.J., 1998, Ecological engineering — the seven-year itch, *Ecological Engineering*, 10, 119.

Mitsch, W.J. and Jørgensen, S.E., Eds., 1989, *Ecological Engineering. An Introduction to Ecotechnology*, John Wiley & Sons, New York.

Odum, H.T., 1962, Man in the ecosystem, in *Proceedings Lockwood Conference on the Suburban Forest and Ecology*, Bull. Conn. Agr. Station 652, Storrs, CT.

Odum, H.T., 1983, *System Ecology*, Wiley Interscience, New York.

Odum, H.T., Siler, W.L., Beyers, R.J., and Armstrong, N., 1963, Experiments with engineering of marine ecosystems, *Publ. Inst. Marine Sci. Uni. Texas*, 9, 374.

Patten, B.C., 1982, Indirect causality in ecosystems: its significance for environmental protection, in Mason, W.T. and Iver, S. (Eds.), *Research on Fish and Wildlife Habitat*, EPA-600/8-82-022, Washington, DC.

Patten, B.C., 1991, Network ecology: indirect determination of the life–environment relationship in ecosystems, in Higashi, M. and Burns, T.P. (Eds.), *Theoretical Studies of Ecosystems: The Network Perspective*, Cambridge, University Press, Cambridge, U.K.

Patten, B.C., Straskraba, M., and Jørgensen, S.E., 1997, Ecosystem emerging, Conservation, *Ecological Modelling*, 96, 221.

Skaarup, R. and Sørensen, T., 1994, *Green Audit of a Farm*, thesis, KVL, Copenhagen.

Straskraba, M., 1984, New ways of eutrophication abatement, in Straskraba, M., Brandl, Z., and Procalova, P., Eds., Hydrobiology and Water Quality of Reservoirs. Acad. Sci., Ceské Budejovice, Czechoslovakia, 37.

Straskraba, M., 1985, *Simulation Models as Tools in Ecotechnology Systems. Analysis and Simulation. Vol. II*, Academic Verlag, Berlin.

Straskraba, M., 1993, Ecotechnology as a new means for environmental management, in *Ecological Engineering*, 2, 4, 311.

# 13

A System Approach to Environmental Management

S.E. Jørgensen

## CONTENTS

## 13.1  Environmental Management of Systems

Management has to focus on systems. A forest is managed, not the trees. An industrial plant is managed not the various production components. Therefore, it is important in a management situation to understand the properties of the system and not necessarily the properties of its components, although they may be reflected in the system properties.

Environmental management is concerned with systems that have a particularly high complexity. It is therefore of particular interest to understand and apply the system approach in environmental management: to consider the entity and not be "disturbed" by the details. An analytical approach that tries to uncover all the details by analytical methods may be used for detailed problems, but not for a proper understanding of a complex system. The systems of environmental interest are too complex to be analyzed in all their details.

This book illustrates the application of system approaches to all systems of relevance for environmental management. In the 1960s and early 1970s, we attempted to solve pollution problems by dilution and removal processes, but during the last approximately 20 years it has been acknowledged that these methods are important but insufficient. An analysis of the different involved systems has led to the conclusion that many environmental problems are best and most economically solved by system changes, resulting from a system analysis. This volume has given an overview of the various system approaches used beneficially in environmental management.

Chapters 3, 4, and 12 looked into *ecosystems*. Chapter 3 attempts to reveal the system properties of ecosystems, while Chapter 4 presents methods to assess the integrity of an ecosystem, based upon our knowledge of the most important components in the ecosystem. Chapter 12 deals with the possibilities of engineering ecosystems. The idea is that if we know the components, their network, and the functions of the system, we will be able to apply the system in our environmental management plan, either directly, by erecting a similar but artificial ecosystem, or by recovering the ecosystem.

1-56670-337-9/00/$0.00+$.50
© 2000 by CRC Press LLC

241

Chapters 5 and 6 looked into *systems encompassing companies,* including production units. How can we improve our production systems from an environmental point of view? A number of ISO-standards present methods to answer this question. The idea is to look into the internal and external flows of mass and energy to be able to assess the unnecessary losses of material and energy and to propose alternative production steps. The ISO-standards may be considered tools in our effort to propose better production methods from an environmental point of view. Chapter 6 is concerned with cleaner production. The question is: Could we produce our products in a more environmentally friendly way? The approach is based on system-thinking. Cleaner production can be obtained (1) by a change in raw material, (2) by use of good operating practices (which may be revealed easily by use of mass and energy balances to assess unnecessary losses), (3) by technological changes (use of other processes), and (4) by reuse and recycling, which eliminate waste products.

Chapter 7 is concerned with green auditing, which is another word for setting up mass balances for a production system, considering all the external and internal mass and energy flows of interest for the production. The chapter introduces a software that can be applied to perform green auditing. It is applied in the chapter on *agricultural systems.* The example presented is the use of nitrogen in Danish agriculture. It is, however, generally applicable to any system. The qualitative result will probably be the same if the method is used on the agriculture of any other country.

The system in Chapter 8 is a *country (or region) — a geographical entity,* in this case Denmark, but the problems and the discussions about how to provide the basic data to set up environmental statistics for a country or region are the same for all industrialized countries. The conclusion is that the data *can* be provided, but it is not an easy task. As an illustration of the problems the nitrogen balance of Denmark is again presented.

Chapter 9 on life cycle analysis, LCA, looks into *a product as a system.* The approach is holistic, as it is characteristic for all the presented environmental management approaches. The product will move in time and space, but the system boundaries move with the product. The object is the system of processes that the substance or material meets within the spatial and temporal boundaries of the study. Mass and energy balances are widely used tools in LCA, as they are for all environmental systems.

Chapter 10 focused on *a chemical compound as the system.* The elements of the system are the properties of the chemical compounds and their concentrations in various environmental compartments. The properties related to the effects (hazards) on the environments as functions of the concentration are of particular interest, i.e., concentration–effect and dose–effect relationships. This includes the effects on all the components of the environment from the physical and chemical to all or at least most of the biological components. The concentrations in all relevant compartments are of interest, such as groundwater, soil, air, water in lakes, rivers, and coastal zones, fish, birds, various plants, phytoplankton, zooplankton, and mammals. The human exposure is of particular interest.

Chemical analyses can be used to determine these concentrations, but mass balances are also used to determine them, often by use of models, due to the high complexity of the systems involved. A combination of chemical analyses and models is most favorable, because the model is used to determine where and when to take samples, and later is used to make simulations after the model has been calibrated and validated by use of the analytical results. The final result of the environmental risk assessment is an answer to the question: How risky is the application of the chemical compound? Can the risk be accepted? If not, what do we do to reduce the risk?

Chapter 11 considered *all chemical compounds,* or at least the 100,000 we are using in such an amount that they may threaten the environment. It attempts to answer the question: Is there a relationship between the structures of chemical compounds and their properties?

To the extent that the answer is yes, it is valuable for the system approaches to know and utilize these relationships.

It is clear from this short overview of the content of the previous 12 chapters that the various system approaches have many features in common, primarily the considerations of entities (systems), the use of holism, and the frequent use of mass and energy balances, often in the form of models. System thinking is different from the classical approach in which analyses are the most important tool. The former considers entities; the latter reveals details. Both approaches are needed, but the system approach has not been used sufficiently to ensure the success of environmental management. The system approach attempts to capture the system properties by synthesizing our detailed knowledge about the system, but without the analyses we have no detailed knowledge to synthesize. On the other hand, if we look only into details, we cannot get an overview and can therefore not identify the roots of the problem. We might easily solve a problem in one part of the system, but create a more serious problem elsewhere.

## 13.2   Integrated Environmental Management

The environmental problems have not been solved even in the countries mastering the most advanced environmental technology and with a willingness to pay the price for an acceptable environmental quality. Trillions of dollars have been invested in solutions for environmental problems, but the problems are still far from being solved, and they will still be major issues in society for the next 30 or 40 years. Some problems have, of course, been solved. For instance, the air quality in many cities is much better today than two or three decades ago. We would, however, have expected that all the problems would have been solved today, considering the enormous investments, if we were asked about it 25 or 30 years ago.

We also touched on this problem in the Introduction, but we shall discuss this issue again here, now that we have an overview of the previous 12 chapters. It is clear that the various system approaches tackle different types of problems in different systems. The recovery of ecosystems is different from production of a more environmentally friendly product, but we need to solve both problems to be able to offer a complete solution to the entire spectrum of environmental problems.

There are two explanations of why we have not yet solved the environmental problems in spite of the enormous investments:

1. The problems were underestimated 25 to 30 years ago. We didn't realize that diffuse pollution was a problem; we didn't realize that the use of chemicals was associated with so many problems, and we didn't realize that global environmental problems would come so much into focus. The underestimation was not only the number of problems, but also the spectrum of problems. Today we acknowledge that the problems are embedded in each production step, in each application of products, and in (almost) every activity of modern society.

2. In addition to the number and size of environmental problems, we also underestimated their complexity. Years ago, we thought all the problems were end-of-the-pipeline problems which could be solved by environmental technology, but many environmental problems are diffuse pollution problems and problems

related to our application of certain products. In addition, many end-of-the-pipeline problems are so expensive to solve, that it has evoked alternatives, for instance the use of cleaner technology, to ensure an economically feasible solution.

Given the enormous complexity of environmental problems and of the involved systems, what can we do to solve them properly? The answer is to use the entire spectrum of methods. We need environmental technology, but without the use of the entire spectrum of system approaches presented in this volume together with environmental modeling, we cannot solve the problems. Only by understanding the reactions and functions of ecosystems to various forms of pollution, only by understanding the relationship between all our activities in the society and the pollution problems, and only by understanding

1. The fate
2. The life cycle
3. The effects on the environments

of our many products, we can come up with a complete, realistic, and economically feasible solution to environmental problems.

These considerations make environmental management a very difficult discipline, because it requires a profound knowledge of a wide spectrum of approaches, tools, concepts, and ideas. One person can hardly be a specialist in all the system approaches presented here, and on environmental modeling, environmental technology, and environmental economy. Therefore, a team of specialists is needed to solve the environmental management problems of today, maybe in close cooperation with a generalist. It is, under all circumstances, important that all team members know at least superficially the field of the other specialists to be able to obtain the right expertise level of the environmental discussion among the team members. This volume may be considered a general knowledge base for such teams. Specialists in each of the various system approaches have to have a much wider knowledge in combination with practical experience.

The conclusion is that not only do we need system approaches in addition to the more classical approaches, which implies a more general, synthesizing, and holistic view of the environmental problems, but we also need to integrate all these approaches with modeling, environmental technology, and environmental economy to be able to come up with environmental management plans that can solve the complex environmental problems in a sustainable manner. Proper environmental management requires a *simultaneous* application of the entire spectrum of relevant approaches to be able to come up with the most economical–ecological solutions to the problems.

# Glossary

*Acceptable daily intake — ADI* — is the estimate of the amount of a considered substance in food or drinking water which can be ingested over a lifetime by humans without appreciable health risk. ADI is normally used for food additives. The applied unit is usually mg/kg body weight.

*Adaptation* — is the response of an organism to changing environmental conditions.

*Adverse effect* — is the change in morphology, physiology, growth, biochemistry, and/or development of lifespan of an organism which results in impairment of the functional capacity or impairment of the capacity to compensate for additional harmful effects of other environmental influences.

*Bioconcentration factor — BCF* — is the ratio of the examined chemical substance concentration in the test organism and the concentration in the test medium, water (or air), at steady state.

*Biological half life (t 1/2)* — is the time needed to reduce the concentration of a chemical in environmental compartments or organisms to half the initial concentration by various biological processes (biodegradation, metabolism, or growth).

*Biomagnification factor — BMF* — is a measure of the tendency of a compound to be taken up through the food. It is the concentration of a chemical compound in a living organism divided by the concentration of the chemical compound in the food at steady state.

*Carcinogenicity* — is the development of cancer. Any chemical which can cause cancer is said to be *carcinogenic.*

*Critical range* — is the range of concentrations in mg/l below which all fish lived for 24 h and above which all died. Mortality is expressed as a fraction indicating the death rate (e.g., ¾).

*DOC* — is the abbreviation for dissolved organic carbon.

*HCp* — is the hazardous concentration for p% of the species, derived by a statistical extrapolation procedure.

*ICp* — is the inhibiting concentration needed to produce an inhibiting effect of p%.

*$LC_{50}$ (lethal concentration fifty)* — is a calculated concentration which, when administered by the respiratory route, is expected to kill 50% of the population of experimental animals during an exposure of a specified duration. Ambient concentration is expressed in milligrams per liter.

*$LC_n$ (lethal concentration n)* — is a calculated concentration which, when administered by the respiratory route, is expected to kill n% of the population of experimental animals during an exposure of a specified duration. Ambient concentration is expressed in milligrams per liter.

*$LD_{50}$ (lethal dose n)* — is a calculated dose of a chemical substance which is expected to kill 50% of a population of experimental animals exposed through a route other than respiration. Dose concentration is expressed in milligrams per kilogram of body weight. The $LD_{50}$ has often been used to classify toxicity between chemical compounds. The following classification may be used — oral $LD_{50}$ to rat, expressed in mg/kg body weight: highly toxic < 25; toxic > 25 and < 200; harmful > 200 and < 2000.

*$LD_n$ (lethal dose fifty)* — a calculated dose of a chemical substance which is expected to kill n% of a population of experimental animals exposed through a route other than respiration. Dose concentration is expressed in milligrams per kilogram of body weight.

*LOEC* — is the abbreviation for *lowest observed effect concentration.* The LOEC is generally reserved for sublethal effects but can in principle also be used for mortality, which is usually the most sensitive effect observed.

*MAC (maximum allowable concentration)* — is a value in accordance with environmental legislation. Often dependent on time. This relation may be expressed as follows: $\log C = 1.8 - 0.7 \log t + 0.068 \log t^2$, where C is the MAC (in mg/m³), and t is the time (in hours).

*NC (narcotic concentration)* — $NC_{50}$ — is the median narcotic concentration.

*No effect level* — *NEL* — implies that the animals remained in good condition. In most experiments, blood and urine tests were made. The urine tests included specific gravity, pH, reducing sugars, bilirubin, and protein. The blood tests included hemoglobin concentration, packed cell volume, mean corpuscular Hb content, a white and differential cell count, clotting function, and the concentration of urea, sodium, and potassium. Control tests for hematological examination were made on the group of animals before exposure. No effect level in this context means: no toxic signs; autopsy — organs normal; blood and urine tests — (if made) normal.

*No observed effect concentration (level)* — *NOEC (NOEL)* — is defined as the highest concentration (level) of a test chemical substance to which the organisms are exposed that does not cause any observed and statistically significant adverse effects on the organisms compared with the controls.

*POM* — is the abbreviation for *particulate organic matter.*

*Predicted environmental concentrations* — *PEC* — is the concentration of a chemical in the environment, calculated primary by use of models on the basis of available information on its properties and application pattern.

*Predicted no effect concentration* — *PNEC* — is the environmental concentration below which it is probable that an unacceptable effect will not occur, according to predictions.

*Predicted no effect level* — *PNEL* — is the maximum level expressed as dose or concentration which on the basis of our present knowledge is likely to be tolerated by a considered organism without producing any adverse effect.

*Principal component analysis* — *PCA* — is a multivariate technique to derive a set of orthogonal parameters from a large number of properties.

*Quantitative structure activity relationship* — *QSAR* — covers the relationships between physical and/or chemical properties of substances and their ability to cause a particular effect or enter into certain processes.

*Retrospective risk assessment* — is a risk assessment performed for hazards that began in the past.

*Risk-benefit analysis* — is the process of setting up the balance of risks and benefits of a proposed risk-reducing action.

*Risk quotient* — is the PEC/PNEC ratio.

*Teratogenesis* — is the capacity of a substance to cause defects in embryonic and fetal development. Any chemical which can cause these defects is said to be teratogenic.

*The median tolerance (TLm)* — this term has been accepted by most biologists to designate the concentration of toxicant or substance at which m% of the test organisms survive. In some cases and for certain special reasons, $TL_{10}$ or $TL_{90}$ might be used. The $TL_{90}$ might be requested by a conservation agency negotiating with an industry in an area where an important fishery exists, and where the agency wants to establish waste concentrations that will definitely not harm the fish. The $TL_{10}$ might be requested by a conservation agency which is buying toxicants designed to remove undesirable species of fish from fishing lakes.

*Threshold-effect concentration* — *TEC* — is the concentration calculated as the geometric mean of NOEC and LOEC. It is equivalent to MATC, *maximum acceptable toxicant concentration.*

*Threshold limit value* — *TLV* — is the concentration in air of a chemical to which most workers can be exposed daily without adverse effects, according to the current knowledge.

*TOC* — is defined as the *total organic carbon* expressed, for example, as kg organic carbon/kg solid. Organic carbon can often be estimated as 50 to 60% of all organic matter.

*Tolerable daily intake* — *TDI* — is the acceptable daily intake established by the European Committee for food. It is expressed as mg/person, assuming a body weight of 60 kg. TDI is normally used for food contaminants, unlike ADI. (See also *ADI.*)

*Toxicity equivalency factor* — *TEF* — is a factor used in risk assessment to estimate the toxicity of a complex mixture of compounds.

*Ultimate median tolerance limit* — *UMTL* — is the concentration of a chemical at which acute toxicity ceases.

*Xenobiotic* — is a man-made chemical not produced in nature and not considered a normal constituent component of a specified biological system.

# Index

247